First Grade Math Workbook

By Greg Sherman

Home School Brew Press

www.HomeSchoolBrew.com

Cover Image © SP-PIC - Fotolia.com

Table of Contents

Disclaimer

This book was developed for parents and students of no particular state; while it is based on common core standards, it is always best to check with your state board to see what will be included on testing.

About Us

Homeschool Brew was started for one simple reason: to make affordable Homeschooling books! When we began looking into homeschooling our own children, we were astonished at the cost of curriculum. Nobody ever said homeschool was easy, but we didn't know that the cost to get materials would leave us broke.

We began partnering with educators and parents to start producing the same kind of quality content that you expect in expensive books...but at a price anyone can afford.

We are still in our infancy stages, but we will be adding more books every month. We value your feedback, so if you have any comments about what you like or how we can do better, then please let us know!

To add your name to our mailing list, go here: http://www.homeschoolbrew.com/mailing-list.html

Algebra

Teaching young children mathematical concepts, such as algebraic thinking, requires that we as teachers and parents make it interesting and motivational. Math is a higher brain function and requires concentration and attention to detail. Algebra is built on sequential steps that are learned and implanted in a student's long term memory. We start with addition, then build on that with subtraction, then build on that with the next concept, and so on. There is an old saying, "New knowledge builds on top of old knowledge." This is especially true with math.

Math requires the student to think, to pay attention, and to repeat. Being distracted by other children, television, and other activities hinders the learning process, especially with mathematical equations. Math is learned by concentrated repetition and practice; there aren't any short cuts. It takes repetition, consistency, effort, and tenacity, which are all traits that will serve the student well in other areas of his/her life. Practice, practice, practice!

Algebra in its simplest form is an equation formulated on basic arithmetic (addition, subtraction, multiplication, and division). For example, if a student knows that 8+2=10 and 2+8=10, then it is easy for them to understand that 10-2=8 and 10-8=2. Then introduce the same problem with blank spaces added: 10-__=8 and 10-___=2. The next step is to add an "x" where the blank spaces were previously: 10-x=8 and 10-x=2. It is only a small jump for the brain to determine that "x" in 10-x=8 is 2 and that "x" in 8+x=10 is 2. As we teach young students basic math, we need to substitute "x" equations into the problems to ease them into algebraic thinking. By teaching the student to make these small transitions from basic arithmetic to algebraic thinking, the child builds an understanding of algebra that will influence how he/she sees math in the succeeding years.

When we are teaching algebra and mathematic principles, the student must commit to memorization the basics so that this information is stored in the long term memory and can be retrieved at a moment's notice when needed for more complex mathematical thinking. As a part of this process, we need to familiarize students not only with the basic principles of arithmetic, but also with the inversion and variations of each particular problem. As in the example given earlier, the student must instantly know that 8+2=10 is exactly the same a 2+8=10. When this information is rote, the jump to algebraic thinking is smooth and easy.

Adding double digits, such as 10+12=22 is easier when the student has learned the basic one number addition combinations by rote. Then we simply ask them to add each column of numbers to get the total. Once they are comfortable with this, we can add the principle of "carrying" a number to the next column, such as 98+37=135. The student would add the 7+8 which is 15 and carry the 1 to the 9+3, which becomes 1+9+3=13.

The math standards encourage students to look at arithmetic in different ways. Now, let's take the last example of 98+37=135 and look at it with different *algorithms*. If we know that 98 is 2 less than 100, we could also approach the problem by taking one of two other approaches:

1. 100 plus 37 is 137, but since the original number was 98, 2 less than 100, we would remove 2 from the 37 making it 35. So we could see at a glance without having to do the column addition that the answer is 135.

2. Or, since we are thinking 100 and it is 2 more than 98, we could remove 2 from 37 making it 35 and add that number to 100, arriving at the answer of 135.

This way of thinking teaches children how things fit together logically. From arithmetic grows algebra and from algebra grows advanced mathematical concepts. The biggest hurtle to learning algebra is that students are weak in basic arithmetic. We need to stress the basics and encourage the kids to memorize single numbers addition, single numbers subtraction, single numbers division, and multiplication tables.

Algebra Worksheet

$$\begin{array}{cccc}
6 & 13 & 84 & 4 \\
+4 & +6 & +11 & +1 \\
\hline
\end{array}$$

$$\begin{array}{ccc}
33 & 91 & 2 \\
+12 & +7 & +0 \\
\hline
\end{array}$$

$5 + 5 = \underline{\hspace{1cm}}$ $8 + 7 = \underline{\hspace{1cm}}$

$15 + 18 = \underline{\hspace{1cm}}$ $22 + 32 = \underline{\hspace{1cm}}$

$$4 + \underline{\quad} = 7 \qquad \underline{\quad} + 6 = 9$$

$$55 + \underline{\quad} = 61 \qquad 24 + \underline{\quad} = 33$$

Solve for X:

$$5 + X = 10 \qquad 12 + X = 24$$

$$X + 7 = 11 \qquad 45 + X = 55$$

$$X + 5 = 15 \qquad 47 + X = 74$$

$$4 + X = 21 \qquad 9 + X = 32$$

3	8	5	4
X4	X1	X3	X6

6	9	7
X3	X7	X0

$$5 \times 7 = \underline{\hspace{1cm}} \qquad 8 \times 2 = \underline{\hspace{1cm}}$$

$$12 \times 12 = \underline{\hspace{1cm}} \qquad 9 + 8 + \underline{\hspace{1cm}}$$

Solve for X:

$$5 X = 10 \qquad 7 X = 21$$

$$9 X = 81 \qquad 11 X = 22$$

$$8 X = 24 \qquad 6 X = 0$$

$$9 X = 9 \qquad 10 X = 50$$

Challenge Questions: $\qquad \dfrac{32}{X} = 4 \qquad\qquad \dfrac{56}{X} = 8$

Capacity

Teaching capacity refers to the volume (or capacity) of a three-dimensional area. There are two types of measures for capacity: solids and liquids. When teaching about capacity, the student must think about two possibilities: (1) How much liquid (or solids) can an object hold in fluid (or dense) measurements? (2) How much space does a three-dimensional object occupy? This is a fun principle to learn because it is hands-on. Everyone loves to play with stuff, right? In teaching we call these items manipulatives.

Capacity Worksheet

Show the student six different sized containers: (Mix the containers up after each question)

1.   Which container has the largest capacity?

2.   Place the containers in order from smallest to largest.

3.   Which container has the smallest capacity?

4.   Place the containers in order from the largest to the smallest.

Definitions: (Write answers on separate piece of paper.)

5.   Who is Archimedes?

6.   What did Archimedes invent?

7. What does displaced mean?

8. What does estimate mean?

9. What does volume mean?

10. What does capacity mean?

11. Which has the largest capacity? Circle the correct item.

 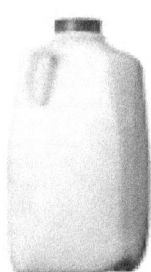

12. Which takes up the smallest volume? Circle the correct item.

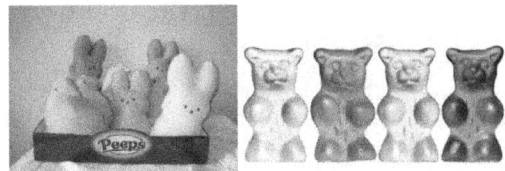

13. Which has the smallest capacity? Circle the correct item.

14. Which container will hold the most jelly beans?

15. Which shown container will hold the most water?

16. Which shown container will hold the least amount of air?

Write answers to questions below on separate piece of paper.

17.   When you estimate the number of jelly beans in a container, how do you determine the exact amount of jelly beans that are actually in the container?

18.   Make up your own problem to explain capacity and show your parent or teacher how to do it.

19.   Using a measuring cup, pour ¼ cup of water into a cup. Add ¾ cups of water to that amount. How much water is now in the cup?

20.   Take a cup of rice. Remove ¼ cup of rice from the cup of rice. How much rice is left in the cup?

Fractions

Teaching fractions can be fun. It is dividing whole items into parts. In its simplest terms, fractions are counting part of something. We write these parts of something as follows: or to put it another way:

The counted part of something
Over the total parts of something

The number of parts chosen
Over the total number of parts

We call the total number of parts chosen the numerator and the total number of parts the denominator. It is important to memorize these terms and to understand their meaning. It would look like this in diagram form:

Numerator
Denominator

In the example of 1/2, 1 is the numerator and 2 is the denominator. In the example of 5/8, 5 is the numerator and 8 is the denominator.

Tell the student about a pizza party. Let's pretend we are ordering an extra large pizza. Since there are two of us, there should be plenty to eat. We get the pizza, notice that it was precut into eight pieces, and chow down on our slices of the delicious pizza. When we are finished, we notice that we each had eaten two slices of pizza and there were four pieces of pizza left. To state this in a fraction, we use a line to divide the two parts. The fraction would look like this:

4 = number of slices that we ate
8 = number of slices in the original uneaten pizza

So, we ate four of the original eight slices, which means we ate half of the pizza. Fractions are the breaking of a whole item into parts. In this case we split the pizza into 8 parts. We ate 4 of those parts. We can explain the amount that is left with fractions. Half of the pizza is left.

Fractions Worksheet

1. Write the parts of these circles in fractions:

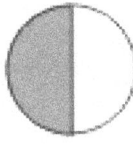   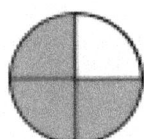   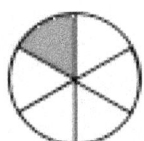   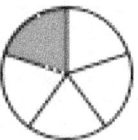   

_____   _____   _____   _____   _____

2. Divide the circles below into the fractions as shown: (Color in the numerators)

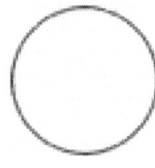   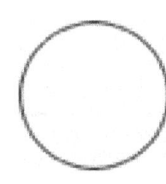   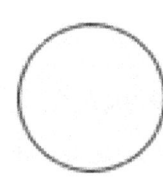   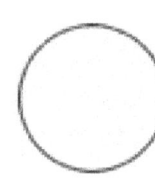   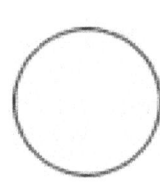

1/2   3/5   4/8   5/6   1/3

3. Color in the numerators as indicated by the fractions below:

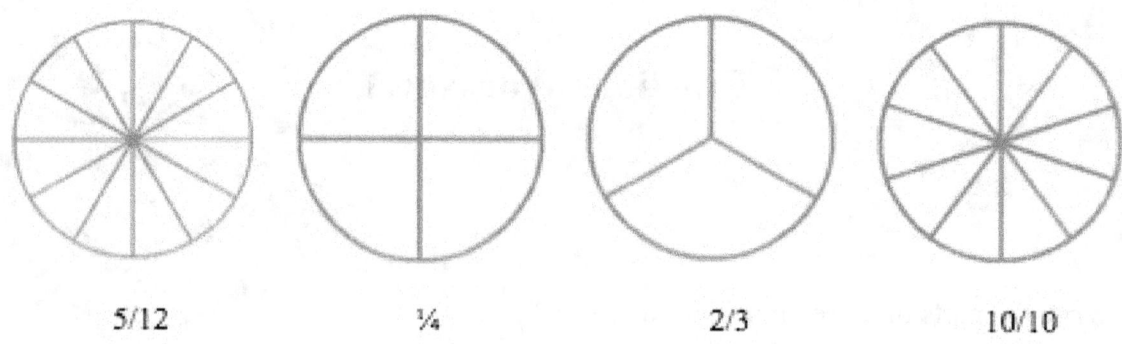

5/12   ¼   2/3   10/10

4.   What is a numerator? _____

5.   What is a denominator? _____

6.   What is a fraction? _____

7.   In the fraction 1/5, what is the numerator? _____

8.   In the fraction 12/13, what is the denominator? _____

9.   What is the number that represents 4/4? _____

10. In the fraction 12/12, what is the numerator? _____

11. In the fraction 9/10, what is the denominator? _____

12. What fraction describes these circles?

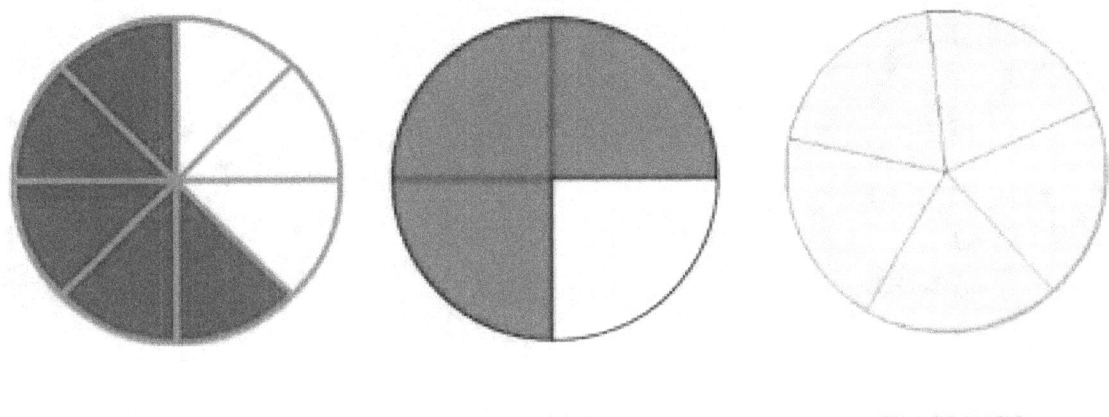

_____   _____   _____

13.   What is the numerator for these circles?

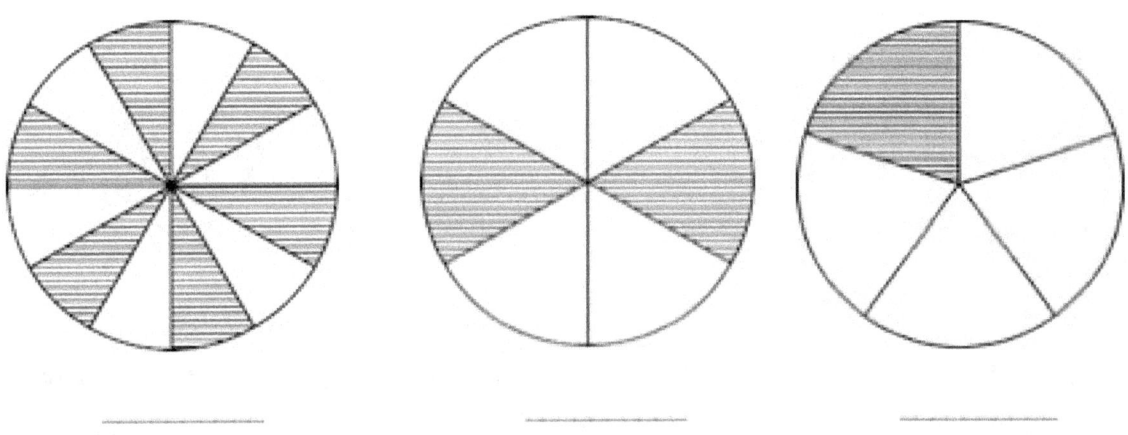

_____   _____   _____

14.   What is the denominator for these circles?

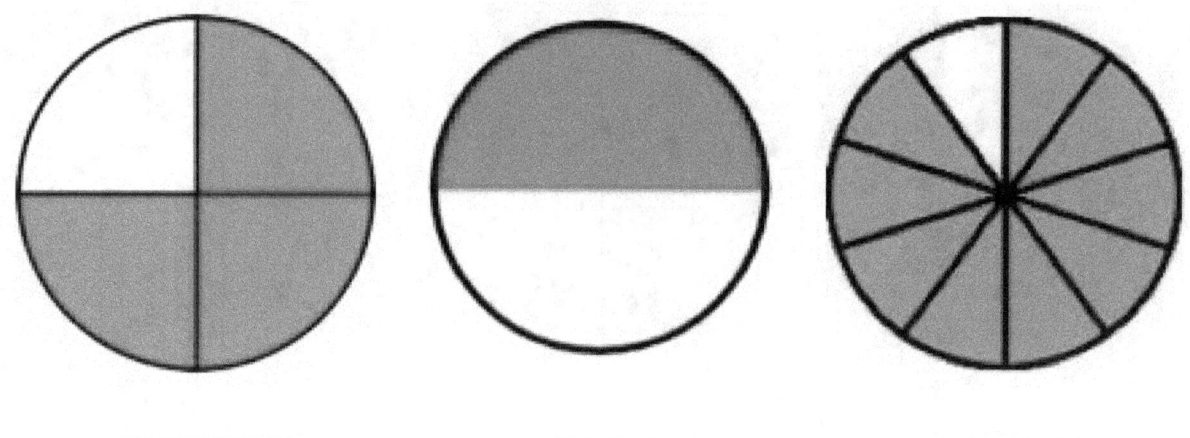

Answer the following questions with the correct fraction:

15. If you have a pizza with 6 slices and you eat 1 slice, you have _____ of the pizza left over.

16. If you have a cookie and you eat half of it, you have _____ of it left over.

17. If a pie has six slices and you eat two of the slices, you have _____ of the pie left over.

18. Draw a circle that represents the fraction 1/2.

19. Draw a circle that represents the fraction 3/3.

20. Draw a circle that represents the fraction ¼.

Length

Length is a measurement that tells us how long something is. There are rulers, tape measures, and protractors that allow us to measure the length of an item. Those measuring devices can be read in inches or centimeters. The metric system is called the System of International Units. The system commonly used in the United States is called the United States Customary Units. For the activities in this unit, a ruler is needed with inches on one side and metric measurements on the other side.

There are many ways to measure things without rulers or tape measures. Have you ever seen someone walk across a room with even steps to measure the size of the room? Or if you can't find a ruler, you can take an object, like a 3" X 5" card and use it to measure the approximate size of a table. We are going to experiment with different ways to measure things in this first activity.

Length Worksheet

1. What is the length in inches of your right shoe? _____

2. What is the length in inches of the pencil you are using? _____

3. What is the length in inches of your ruler? _____

4. What is the length in centimeters of your ruler? _____

5. Looking at your ruler, how many inches are in a foot? _____

6. Mark on the ruler below a length of 2 and 1/4 inches:

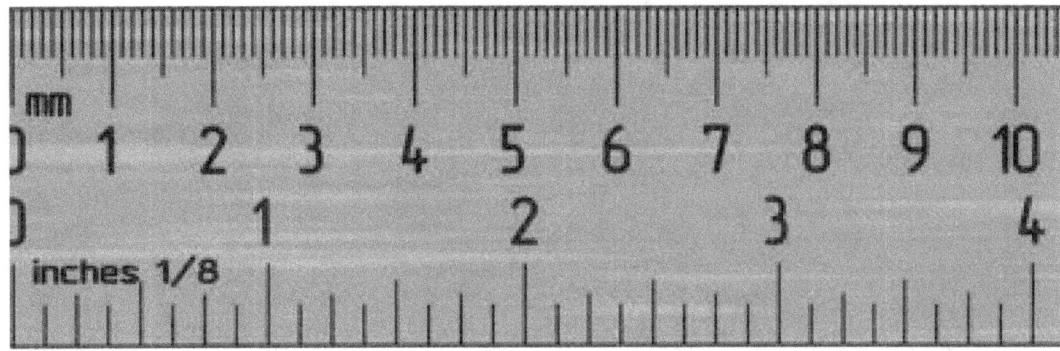

7. Mark on the ruler below a length of 7 centimeters:

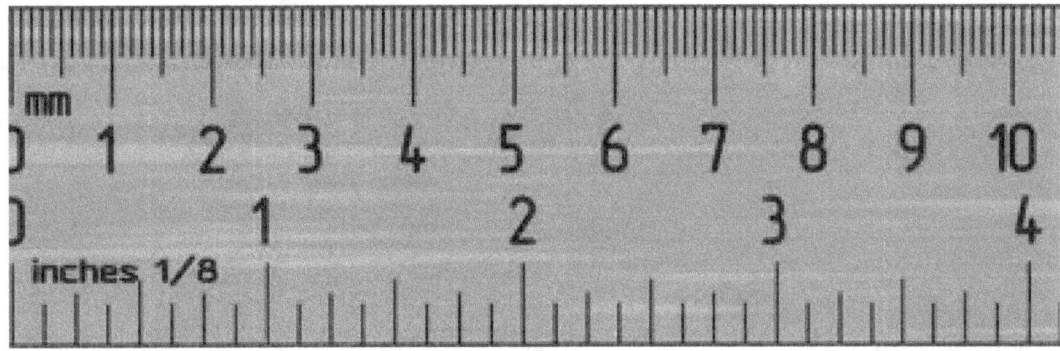

8. Mark on the ruler below a length of 3 and 1/2 inches:

9. Mark on the ruler below a length of 4 and 1/2 centimeters:

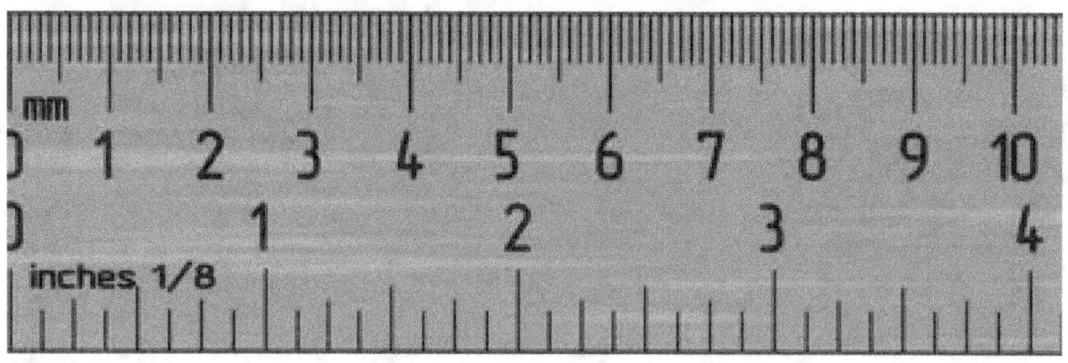

10. Mark on the ruler below a length of 3 and 3/4 inches:

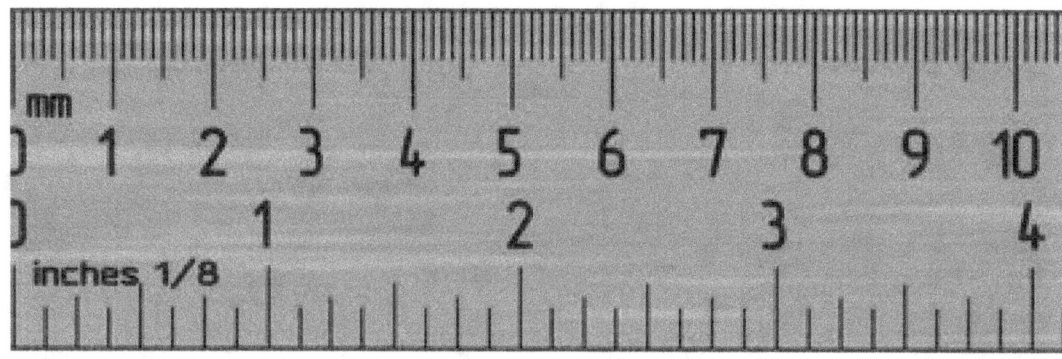

11. Looking at the ruler above, approximately how many centimeters are in 2 inches?

_____

12. Looking at the ruler above, approximately how many centimeters are in 4 inches?

_____

13. What is a definition of length? _____

_____

14. Describe two ways you can measure things:

_____

_____

15.   Briefly explain the problem the King needed to solve in the book "How Big Is a Foot":

_____

_____

16.   What item did you measure that was 1" in length? _____

17.   What item did you measure that was 1 foot in length? _____

18.   Looking at your ruler, how many 1/4 inch lengths are in 1 inch?

_____

19.   Looking at your ruler, how many 1/2 inch lengths are in 1 inch?

_____

20.   Looking at your ruler, how many 1/8 inch lengths are in 1 inch?

_____

Money

"Money makes the world go round ..." even in First Grade! Along with counting and understanding coins, this is the time to add paper money to the mix. Understanding coins involves an awareness of the differences between all of the coins: size, shape, thickness, weight, color, value, and embossed images. By adding the paper money to the mix, the student learns about the monetary value of a combination of paper and coin currency.

For this lesson money will be needed. Real money can be used or images of paper money and coins can be downloaded from the internet and printed on heavy paper and cut out. Even if using fake money, it is a good idea to show the student the real thing so that there is a form of reference. Coins needed are the penny, the nickel, the dime, and the quarter. The paper bills needed are the one dollar bill, the five dollar bill, the ten dollar bill, and the twenty dollar bill.

Show the student all of the coins. Take the time to carefully look at the designs on the front and back of each coin. Read the words imprinted on each coin and discuss the values of each. This thoughtful process should also be done with the paper bills. Look at each bill and talk about the Presidents' photos that appear on each one as well as the other symbols on the front and back of each bill. Point out the security measures on the bills and explain how these features help to prevent counterfeiting. Explain in brief and simple terms about counterfeiting and its impact on the country's economy. Have the student identify the similarities of each paper bill and the noticeable differences. Also explain that the images can change on the coins and paper bills as they are redesigned over time. Point out the features that remain the same so that the coins and bills values are always readily identifiable.

Learning how to count and use currency is an important skill to learn. Explain the meaning of monetary value. The activities in this lesson will help to reinforce these skills.

Money Worksheet

Look at the coins in the middle column. Count up the value of the coins. Write the number value of the coins in the right hand column. The first line is an example.

Ex		**25¢**
1		¢
2		¢
3		¢
4		¢
5		¢
6		¢
7		¢
8		¢

9	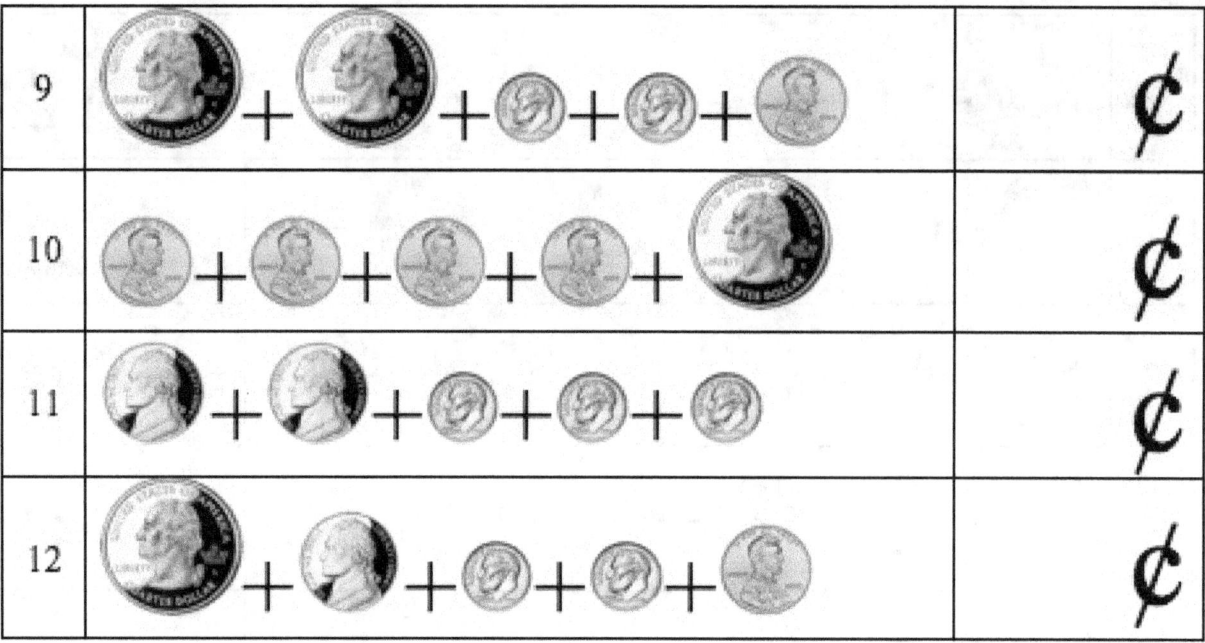	¢
10		¢
11		¢
12		¢

Look at the paper currency in the middle column. Count up the value of the paper currency. Write the number value of the paper currency in the right hand column. The first line is an example.

EX	$1 + $5		$6.00
13	$10 + $20		$
14	$1 + $10		$
15	$5 + $20		$
16	$20 + $1		$

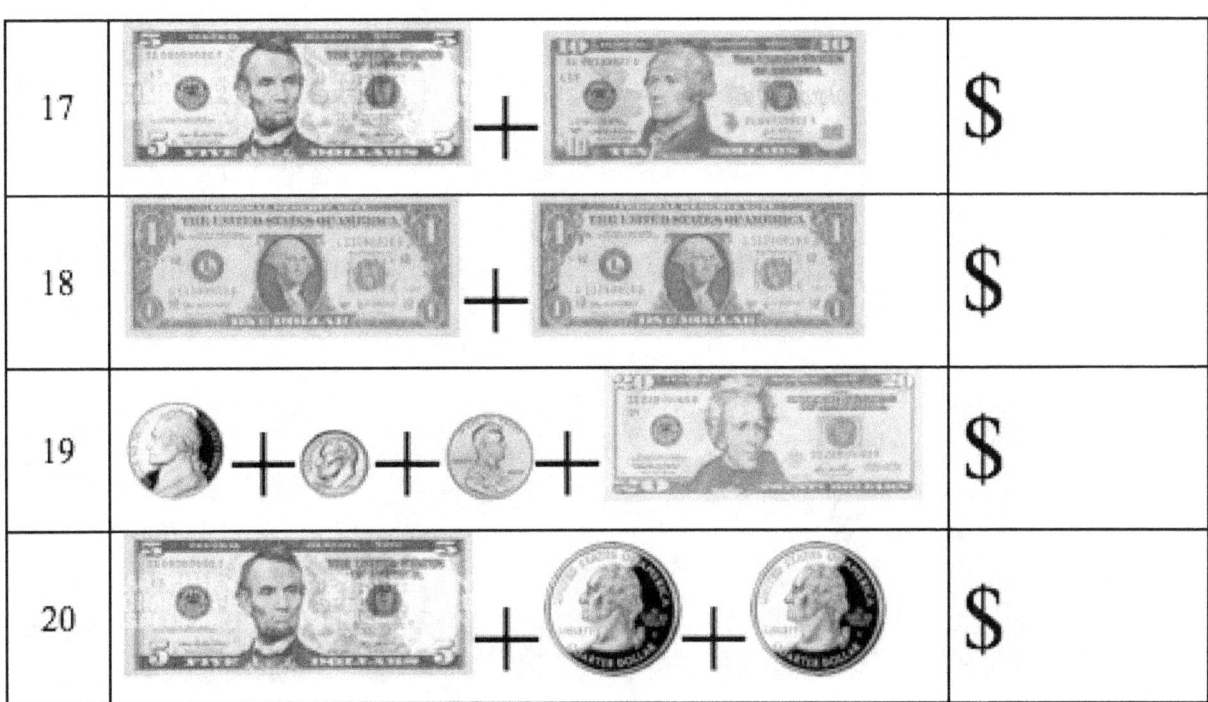

17	$5 + $10		$
18	$1 + $1		$
19	coin + coin + coin + $20		$
20	$5 + quarter + quarter		$

Number

Teaching children number sense improves their ability to explore numbers, count, and understand how much there is of something. It is a way of thinking that is intuitive and flexible. As students are exposed to different situations and activities, they begin to think about numbers in different ways. This helps a student to recognize when he/she is right or wrong and to self correct his/her own errors. These skills enable students to count, compare, estimate, measure, and forecast.

It is important that a student learn not only to count from 1 to 100 in order, but to understand the numbers out of order and to be able to count by 2's, 5's, and 10's to 100. When students become comfortable with numbers, they are able to do arithmetic easier followed by a more complete understanding of algebraic algorithms. As they work with numbers more and more, patterns emerge and enable the students to learn and understand mathematics at a deeper level.

Number Sense Worksheet

1.   Starting in the upper left hand corner and going across one row at a time from top to bottom, number each of the squares from 1 to 100.

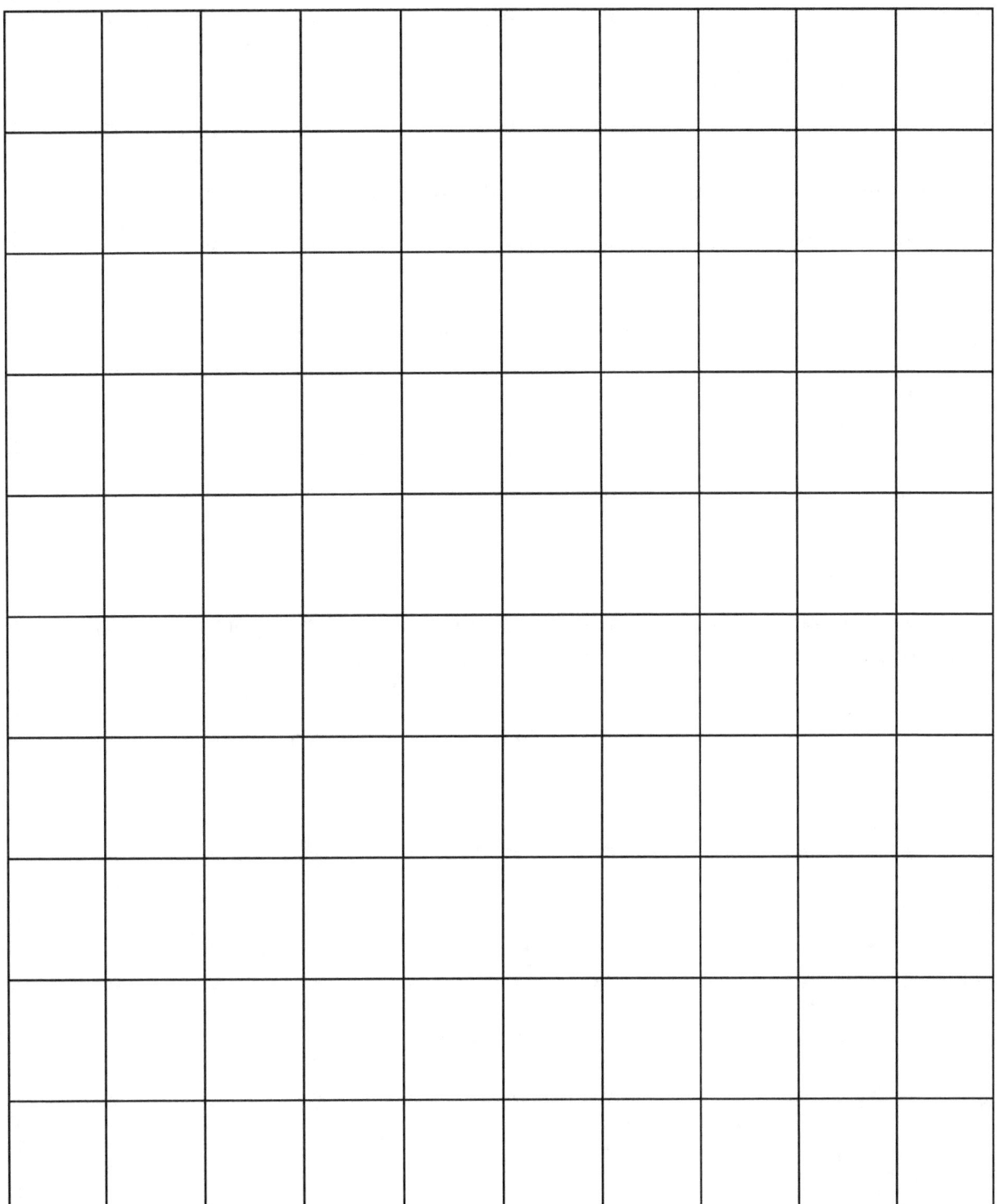

2.   What is a tip for remembering symbols for more than and less than?

_____

3. What are the three comparison symbols and what do they mean?

_____ means _____

_____ means _____

_____ means _____

4. Write the correct comparison symbol in the blank boxes for each set of numbers.   (< = >)

1)	82		5
2)	37		37
3)	1		99
4)	55		21
5)	29		30
6)	99		99

7)	16		10
8)	44		49
9)	68		70
10)	71		17

5. Starting in the upper left hand corner and going across one row at a time from top to bottom, number each of the squares counting by 2's.

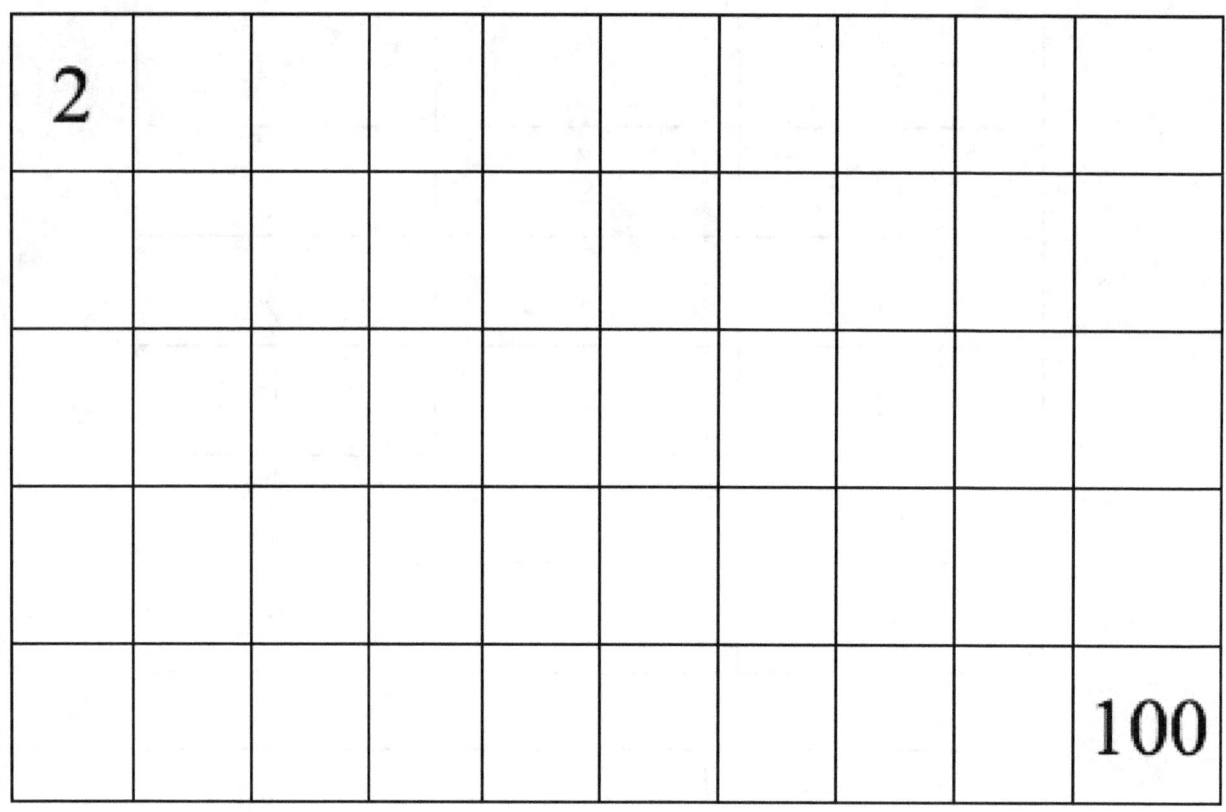

2									
									100

6. Starting in the upper left hand corner and going across one row at a time from top to bottom, number each of the squares counting by 5's from 5 to 100.

7. Starting in the upper left hand corner and going across one row at a time from top to bottom, number each of the squares counting by 10's.

8. Add the numbers for the sums to these problems:

1	2	13	14	15	16	7
8	9	10	11	12	13	1

9. Subtract the numbers to answer these problems:

20	19	18	17	16	15	14
1	0	10	3	11	10	4

10. What is the inverse of this addition problem $11 + 6 = 17$?

_____

11. What is the inverse of this subtraction problem $20 - 11 = 9$?

_____

12. What is the inverse of this addition problem $10 + 17 = 27$?

_____

13. What is the inverse of this subtraction problem $11 - 3 = 8$?

_____

14. What number is ten less than 34? _____

15. What number is ten more than 74? _____

16. What number is ten less than 99? _____

17. What number is ten more than 11? _____

18. What number is one more than 37? _____

19. What number is one less than 83? _____

20. What number is one less than 46? _____

Operations

As a part of the core curriculum, children are being taught skills in operations and algebraic thinking. This method of instruction will help them think in algebraic and abstract terms and will provide a good base for algebra, geometry, trigonometry, and calculus as they advance through the higher grades. Like so many other things in life, algebraic thinking is a mind set or a way of looking at things. When children become familiar with this type of thinking, they can embrace the fun of mathematics.

In this lesson, students will learn to solve word problems and become fluent with the addition and subtraction operations needed to facilitate this problem solving strategy. Students learn how to use an equal sign to determine whether an equation is true or false. Then they progress to an understanding of the commutative and associative properties of addition and subtraction, which leads to an understanding of basic algebraic thinking. These lessons are the first steps to entering the fascinating world of algebra and advanced mathematics.

Operations Worksheet

Using the commutative property of addition, rewrite these problems in a different way.

1. $2 + 8 + 10 = 20$ _____

2. $5 + 5 + 20 = 30$ _____

3. $2 + 2 + 7 = 11$ _____

Using the associative property of addition, rewrite these problems in a different way.

4. $5 + 7 = 12$ _____

5. $8 + 1 = 9$ _____

6, $6 + 4 = 10$ _____

7. What is the meaning of the equal sign? _____

_____

Look at the problems below with an equal sign. Decide if they are true or false. Circle your answer.

8. $8 + 1 = 9$ TRUE   FALSE

9. $2 + 1 + 1 + 4$ TRUE   FALSE

10. $5 = 7$ TRUE   FALSE

11. $3 + 5 = 8$ TRUE   FALSE

12. $6 + 2 = 4 + 5$ TRUE   FALSE

13. $4 + 4 = 7$ TRUE   FALSE

14. $1 + 1 + 1 = 3$ TRUE   FALSE

15. $7 + 3 = 10 - 1$ TRUE   FALSE

16. Write the numbers 10 to 100, counting by tens. _____

_____

17. Write the ODD numbers only from 1 to 20: _____

_____

18. Write the EVEN numbers only from 1 to 20: _____

_____

19. What is the definition of sum? _____

_____

Give an example of a sum: _____

20. Will the sum of 3 + 3 be an even or odd number? _____

Will the sum of 3 + 4 be an even or odd number? _____

Will the sum of 11 + 9 be an even or odd number? _____

Will the sum of 7 + 10 be an even or odd number? _____

Patterns

Teaching students about patterns allows them to analyze different relationships with numbers and groups of numbers and allows them to learn to make predictions. The activities in this lesson will help students learn about growing and repeating patterns. An educator named Robert Wirtz once said, "Arithmetic begins with learning to count by ones, after that, it is a never-ending search for shortcuts to avoid one-by-one counting." Learning to recognize patterns is one of those shortcuts.

A pattern is anything that reoccurs. It can be numbers, designs, or nature. For our purposes, we will be thinking about patterns as they relate to numbers and symbols that are representing numbers. Repeating patterns, just as the name implies, are sequences that never change and can repeat over and over again. Growing patterns, on the other hand, have a starting point and they grow at intervals to infinity. These are shortcuts because they allow us to predict what will come next without going through the slow process step-by-step to figure out the predictions. This lesson will introduce the student to some of these shortcuts.

Patterns Worksheet

1. Describing Patterns: Carefully look at the two diagrams below and answer the questions:

Diagram A

Diagram B

☐ How are the two patterns alike? _____

_____

_____

_____

_____

☐ How are the two patterns different? _____

_____

_____

_____

2. What is a 5-unit sequence? _____

_____

3. What is a repeating pattern? _____

4. Repeating Patterns: Using markers that are the same colors as the 4-unit pattern, color in the remaining boxes, following the pattern exactly until you run out of boxes.

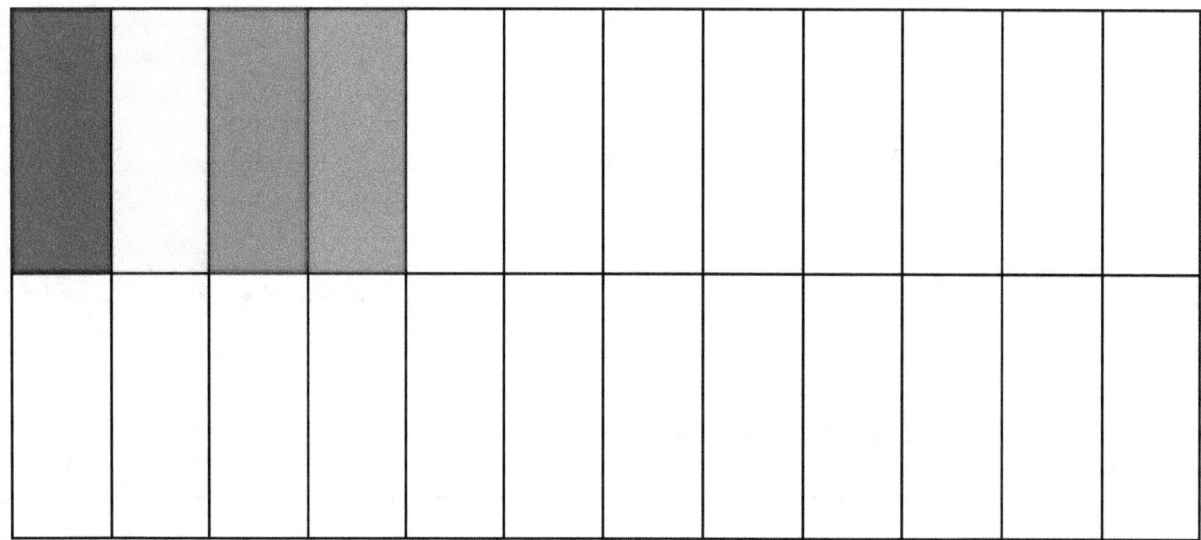

5. Answer the following questions about the Repeating Patterns above:

☐ How many boxes are in the above unit pattern? _____

☐ How many times did you repeat your pattern? _____

☐ Were any column patterns created? _____

Describe them: _____

☐ Were any diagonal patterns created? _____

Describe them: _____

_____

6. Using the following number sequence, repeat this 5-unit pattern sequence 2 more times, making 3 sequences.

13579 _____

7. Look at the growing pattern below and add the next sequence:

8. Look at this growing sequence of circles and add the next sequence:

9. Look at this growing sequence of numbers and add the next sequence:

9 9 7 9 7 5 9 7 5 3

_____

Are the patterns below repeating or growing?

10. _____

11. _____

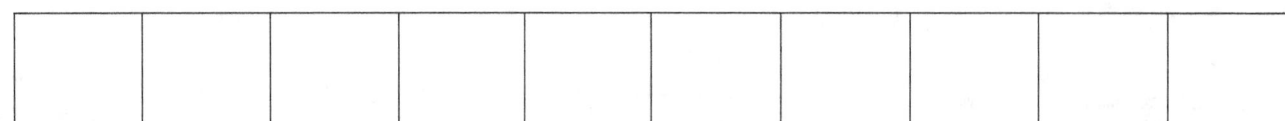

12. _____

13. _____

123A123A123A23A

14. _____

1 1 A 1 A 2 1 A 2 B

15. _____

☐☐✚☐✚☐✛☐✚☐✛☐★☐✚☐✛☐★☐✪☐✚☐✛☐
★☐✪☐☐✚☐✛☐★☐✪☐☐☐

16. What shapes complete the last sequence?

☐☐☐☐☐☐☐☐☐☐☐☐ ___ ___ ___ ___

17. What numbers complete the last sequence?

1 1 2 1 2 3 1 2 3 4 1 2 3 4 5 1 2 3 4 5 6 ___ ___ ___ ___ ___ ___
___

18. Create your own repeating pattern using at least 3 sequences:

19. Create your own growing pattern using numbers and 4 sequences:

20. Create your own growing pattern using shapes in 3 sequences:

Positions

Teaching children positions refers to their spatial sense. They need to be able to clearly recognize locations of objects and numbers to the left, middle, right, up, center, down, top left, center left, bottom left, top center, bottom center, top right, center right, bottom right, inside, outside, under, and over.

One fun activity you can do that will help teach spatial sense is blowing bubbles. When the bubbles float in the air, you can discuss which direction they are floating, where they are landing, how long it takes them to fall and land on an object, and where they finally land and pop.

Another fun activity you can do at home in a safe place is a blindfolded obstacle course. Using pillows and soft items for obstacles, you can blindfold your child and give him/her directions using positions such as you're getting warm, a little to the left, step up, step down, go to the right, turn around, and so on.

Yet another game you can play at home is finding hidden objects. You can direct your child by saying clues like: It is located over the washing machine. It is under the table. It is around the corner. It is behind the sofa. It is under the umbrella. You can point out the lights and ceiling fan are over your head and the rugs, tile, and carpet are below your feet. All of these positions that adults take for granted can be new or unfamiliar to young children.

And of course, and all time favorite is "Simon says." You can do a lot of spatial sense recognition with this cute game. Remember in this game you can only move if the speaker says "Simon says" before the command. "Simon says to step over the pillow." Go around the sofa. "Simon says to jump to the left." Jump to the right. "Simon says to turn to the right and look up."

Positions and spatial sense as it relates to math and school work revolves mostly around printed paper or computer activities. The following activities will help you teach your student positions.

Positions Worksheet

| Pink |
| Purple |
| Green |

Answer these questions about the boxes above.

1. Which box is above the purple box? _____

2. Is the green box above or below the purple box? _____

3. Which box is in the middle? _____

4. On which side is the child?

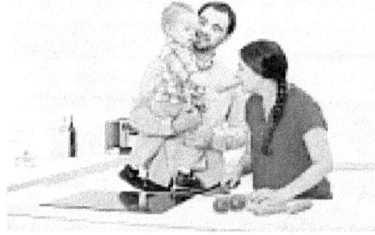

Circle the best answer:

Left   Right

5.   Circle the girl on the right.

6.   Circle the person in the middle

7.   In what position is the woman who is looking at us?

Circle the correct answer:

Left   Center   Right

8.   In what position is the grandmother with the laptop?

Circle the correct answer:

Right   **Left**

Circle the correct answer:

Right   **Left**

9.   Circle the father on the right side.

Put an X on the mother on the left side.

10. Label the hands right and left:

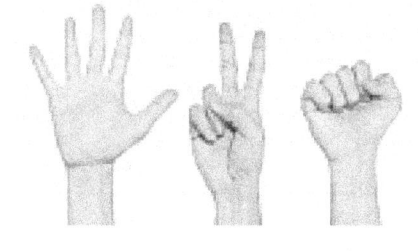

_____   _____

11. Is the yellow star in the middle, on the top or on the bottom?

_____

What is the position of the red star?

_____

12. What does position mean? _____

_____

_____

13. What is on the top of this ice cream?

Cherry

Ice Cream

Bowl

_____

What is on the bottom?

_____

14. Is the square inside, on, or outside the circle?

_____

15. Is the heart inside, on, or outside the square?

_____

16. Is the star inside, on, or outside the Triangle?

_____

Locate the position of the following objects on a 3 X 3 grid:

17. Circle the object in the top right position.

18. Draw a square around the object in the center position.

19. Draw an "X" over the object in the bottom left position.

20. Draw a triangle over the object in the top center position.

Shapes

Teaching children about shapes is the first step in understanding geometry. As children build their knowledge of shapes, open and closed shapes, identifying shapes by name and appearance, shape comparisons, solid versus planar surfaces, and identifying symmetry in objects and nature, they develop their abilities to think in geometric terms. This lesson will focus on this core curriculum skill.

Learning about shapes is also cross disciplinary. These shapes will improve your child's handwriting, improve his/her drawing skills, and develop his/her sequencing and logic skills, which will be used later for such subjects as calculus. We all use an internal knowledge in our every day life when we arrange furniture in our homes, organize our cupboards or our refrigerator, or when we pack our luggage to go on a trip. Knowing the shape and size of items allows us to put them together in organized ways to enhance our life.

Because we are surrounded by shapes, it is easy to introduce children to shapes at home. We can play "Find the Shape" and look for shapes in the design of our home furnishings, our dishes, our automobiles, and in books and magazines. Look for two dimensional and three dimensional shapes and help your child recognize these shapes all around him/her.

To start, focus on these 2D (& 3D) shapes: square (cube), rectangle (cuboid), triangle (pyramid), pentagon, hexagon, circle (sphere), and cylinder

2D vs. 3D? Of course, along with understanding shapes, it is important to understand the difference between two dimensional (2D) and three dimensional (3D) shapes. 2D shapes are basic flat (orthographic in nature) and only have two dimensions – length and width. 3D shapes have volume and have three dimensions – length, width, and depth. So in its simplest terms, a circle is 2D and a ball is 3D. A square is 2D and a cardboard packing box is 3D. A rectangle is 2D and a brick is 3D. A triangle is 2D and the pyramids of Egypt are 3D. Look for objects around your house or out in nature to help your child understand the differences between 2D and 3D.

2D and 3D also relates to surfaces. A two dimensional object is on one plane, or flat, so we refer to it as a planar surface. A three dimensional object is on more than one plane and we refer to it as solid. Children need to understand that a two dimensional surface is planar and that a three dimensional surface is solid. Again, using your home or nature as your guide, find planar and solid objects to reinforce this learning experience.

Shapes Worksheet

1.   Circle the shapes that are two dimensional (2D) below:

2. Circle the shapes that are three dimensional (3D) below:

3. What makes an object 2D (two dimensional)? _____

_____

4. What makes an object 3D (three dimensional)? _____

_____

5. Circle the closed shapes below:

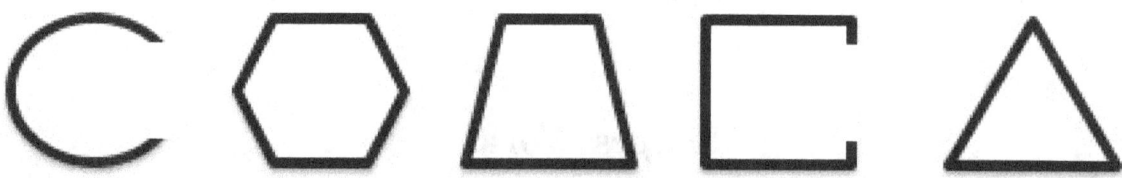

6. Circle the open shapes below:

7. Circle similar shapes:

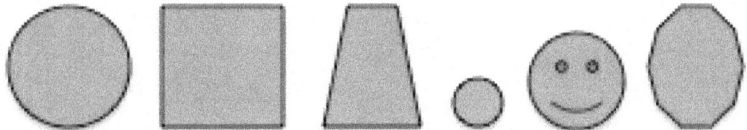

8. Circle dissimilar shapes:

9. Draw a line connecting the name of the shape with the shape:

Square   Circle   Hexagon   Rectangle   Pentagon   Triangle

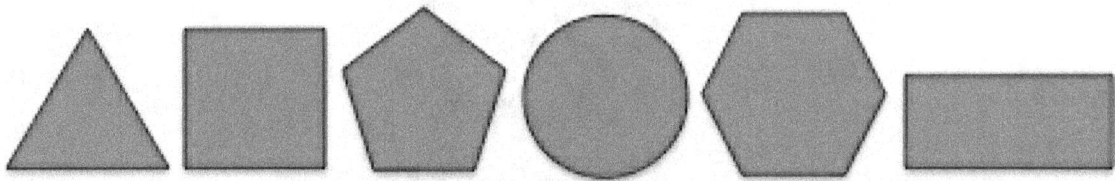

10. Draw a line connecting the name of the shape with the shape:

Sphere	Cylinder	Cube	Cuboid	Pyramid

11. Draw a line connecting the name of the shape to its comparison:

Sphere	Pyramid	Heart	Cylinder	Cube

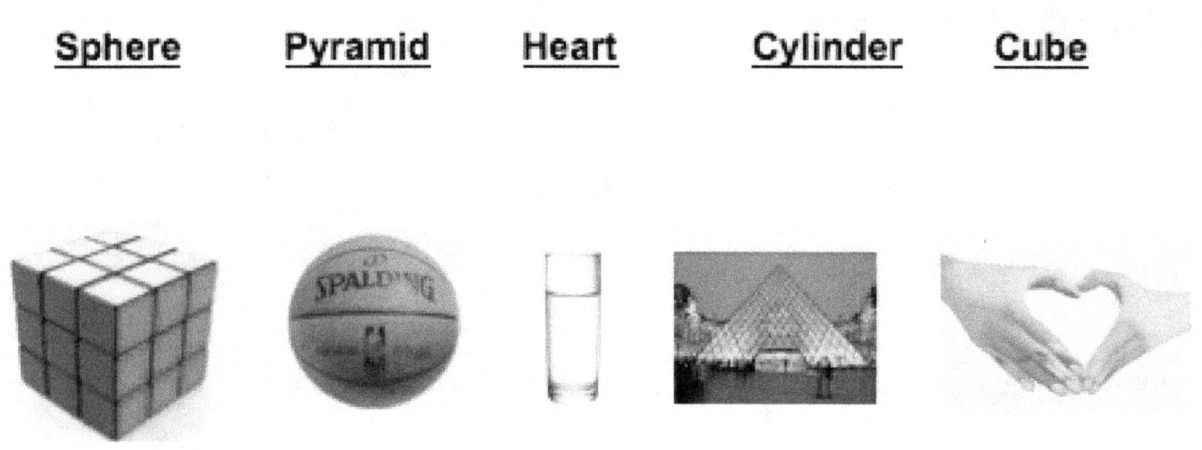

12. What shape looks like this? ✓ the correct answer.

☐ square

☐ pentagon

13. What shape looks like this? ✓ the correct answer.

☐ cuboid

☐ pyramid

14. What shape looks like this? ✓ the correct answer.

☐ rectangle

☐ cylinder

15. Which planar shape relates to this figure?

☐ rectangle

☐ circle

☐ triangle

15. Which planar shape relates to this figure?

☐ pentagon

☐ square

☐ heart

16. Which planar shape relates to these objects?

☐ triangle

☐ hexagon

☐ square

17. Is this object symmetrical?

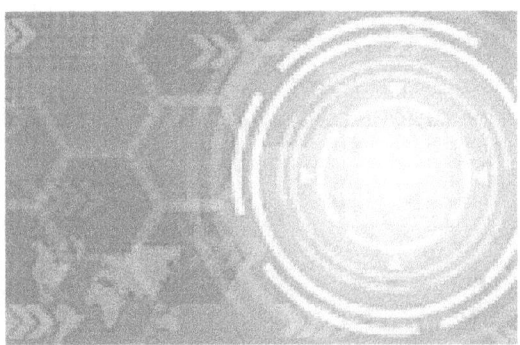

☐ yes

☐ no

18. Is this animal symmetrical?

☐ yes

☐ no

19. Circle the objects that have symmetry:

20. Draw a line through the object below to show symmetrical halves:

Spatial Sense

Teaching children spatial sense covers a lot of territory from math to geography. Not only do we need to learn about our space in the world but also the space of objects and numbers within their worlds. This lesson is designed to be interdisciplinary in nature, connecting your child with not only mathematical concepts, but also concepts of geography and historical context.

Let's first look at our place in the world. What space do we occupy?

Talk about where you are:
In our own solar system
on our planet earth
within a continent
within a country
within a state
in a county
in a city
home
standing or sitting where you are today

Assist your child with looking at a globe and finding your approximate location. Then get a map of your state and find the location of your city. Then go online to Google maps and find the exact location of your home. You might even be able to zoom in on the images and look at your street from the Google satellite. This gives your child a good idea of where he/she is in space and time.

Spatial Sense Worksheet

1.   Fill in the spaces with your personal information:

Who Am I?

My name is _____

The street I live on is _____

My street is in the city of _____

My city is in the county of _____

My county is in the state of _____

My state is in the country of _____

My country is on the continent of _____

My continent is on the planet called _____

Look at a cube shape.

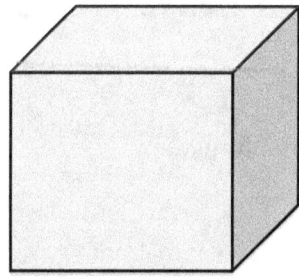

2.   How many sides does it have?

62

_____

3.   What 2D shape makes a cube?

_____

Look at a cylinder shape.

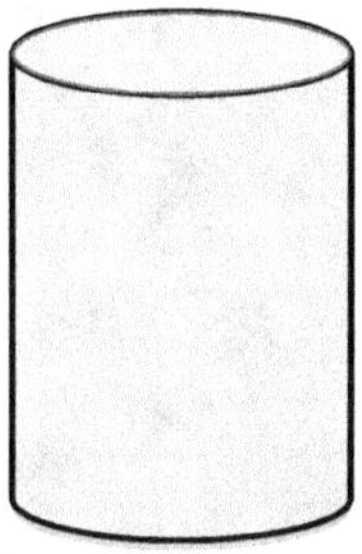

4.   If you cut the cylinder and lie it flat,

      what 2D shape do you have?

      _____

5.   If you want to put a top and bottom on the cylinder, what shape will you need to do that?

      _____

6.   Look at this pyramid. How many <u>triangles</u> make up this pyramid?

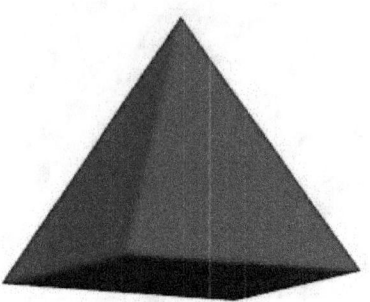

_____

7.   What 2D shape do you need to make a 3D shaped cone?

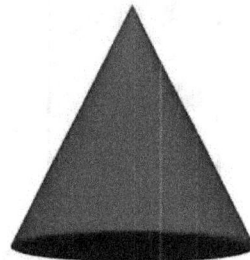

☐   circle

☐   square

☐   half circle

8.   In the cone shape above, what shape is the base of the cone?

_____

Look at this cuboid shape.

9.   How many sides does it have?

_____

10.   What 2D shapes make this cuboid?

_____ and _____

8.   How many squares does it take to make a cube?

   ☐   3

   ☐   4

   ☐   6

9.   How many sides does a closed cone shape have?

   ☐   3

   ☐   2

   ☐   1

10.   How many sides does a closed cylinder shape have?

   ☐   1

   ☐   2

11.   Explain how a box is made from squares. _____

_____

_____

_____

12.   Explain how to make a cone from paper. _____

_____

_____

_____

13.   What 3D shape can you make from   ? _____

14.   What 3D shape can you make from ☐ ? _____

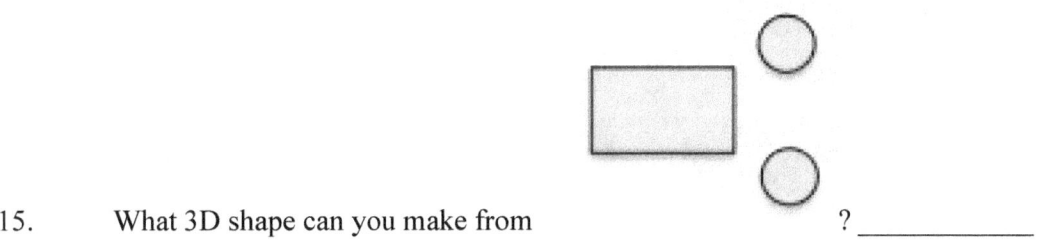

15.   What 3D shape can you make from   ? _____

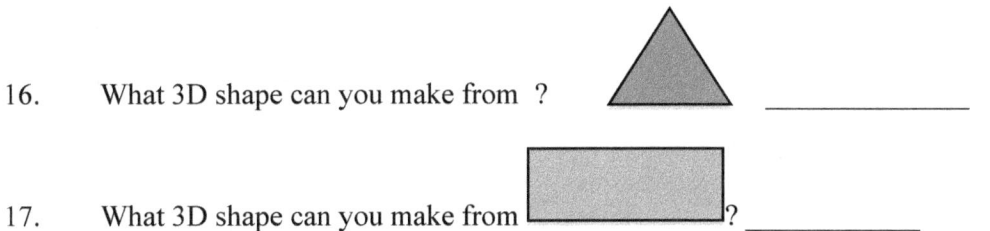

16.   What 3D shape can you make from ? _____

17.   What 3D shape can you make from ? _____

Match up the shapes with similar shapes from a different viewpoint.

18.   Draw a line from the name of the shape to the shape from different viewpoint.

Hexagon   **Pentagon**   **Circle**   **Square**   **Rectangle**

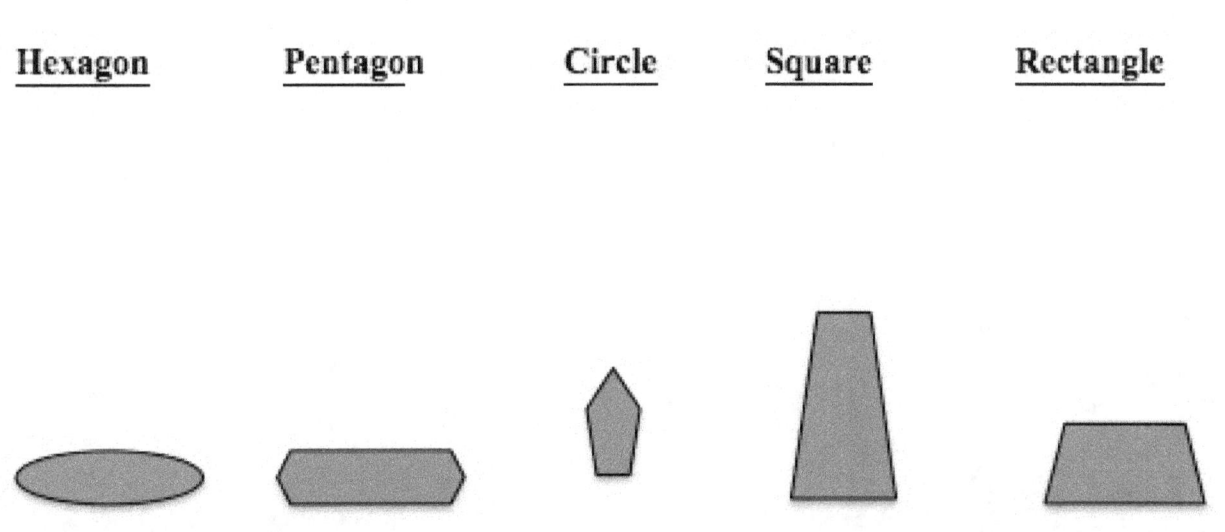

19.   What is your favorite shape? Draw it!

20.   What shape is the yellow portion of this figure?

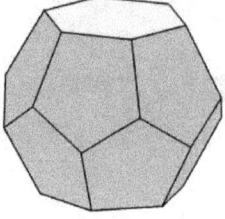

_____

What shapes are the purple portions?

_____

Temperature

Teaching about temperature is one of those interdisciplinary areas that combines mathematics with science. Children will learn the difference between hot and cold, body temperature, hottest and coldest places on earth, how to read a thermometer, and Fahrenheit versus Celsius. Children interact with temperature from the moment they take their first breaths as infants. By exploring the world around them, they can gather and analyze information about their world.

Explain to children that basically temperature is how cold or hot something is at any given moment in time. On our planet, the weather causes the temperature to rise or fall, and the temperatures are affected as well by the rotation of the earth and the earth's proximity to the sun. Water, air, human bodies, animal bodies, food, appliances, and automobiles are just a few things that have temperatures and respond to heat and cold.

What is the coldest place on earth?

Antarctica is the coldest place on earth.

Even though it is all ice and snow, it is the largest desert on the planet. It is also the windiest and highest continent on earth. Because it is so cold and dangerous, few people live and work there.

What is the hottest place on earth?

Death Valley, California, holds the record.

As the hottest place on earth at 134° Fahrenheit recorded in 1913. There are many hot places (Africa, Asia, Australia, and South America to name the most common) with temperatures rising well over 100 degrees Fahrenheit each year.

Temperature Worksheet

1. What is the coldest place on earth?

    ☐ North Pole

    ☐ Antarctica

    ☐ Canada

2. What is temperature? _____

    _____

3. What is a thermometer? _____

    _____

Look at the two thermometers.

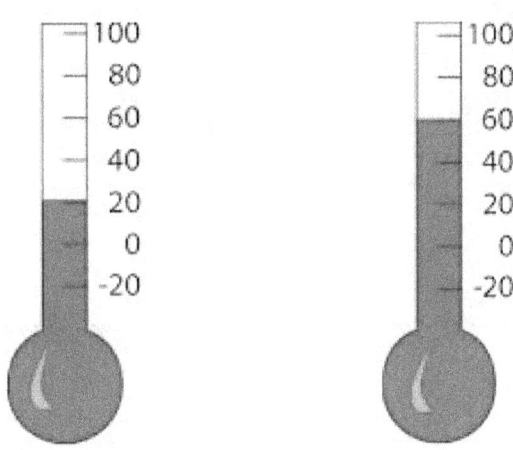

4. What is the temperature in degrees on the thermometer on the left side?

   _____

5. What is the temperature on the thermometer on the right side?

   _____

6. According to this thermometer, it is 74°F.

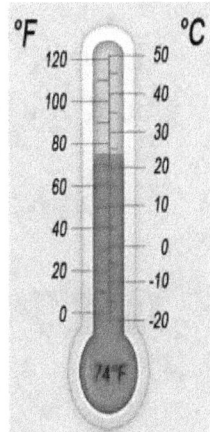

What is the approximate temperature in Celsius?

_____

7. Is this a picture of an analog or digital thermometer?

_____

8. What is the symbol for degrees?   _____

9. What is the symbol for Fahrenheit?   _____

10. What is the symbol for Celsius?   _____

11. Using the correct symbols, how would you write the temperature for 65 degrees Fahrenheit?

_____

12. Using the correct symbols, how would you write the temperature for 30

degrees Celsius?

_____

13. At what Fahrenheit temperature does water boil? Check the correct answer.

☐   150° F

☐   212° F

☐   250° F

13. At what Celsius temperature does water boil?

☐   300° C

☐   500° C

☐   100° C

14. What is the average normal body temperature in Fahrenheit?

☐   96.8° F

☐   97.0° F

☐   98.6° F

15. What is the average normal body temperature in Celsius?

☐   37.0° C

☐   36.6° C

☐   35.8° C

16. What is the hottest time of the day? _____

17. How does the sun affect temperatures? _____

_____

_____

18. How does rain affect temperatures and why? _____

_____

_____

19. Mark this thermometer to show 70° F.

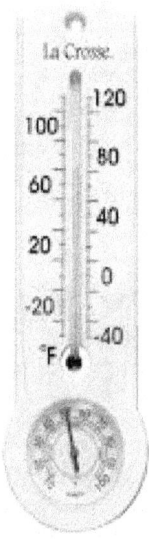

20. What is the highest temperature measurement available on this thermometer?

      _____

     What is the lowest temperature measurement available on this thermometer?

      _____

Time

Learning about time starts when your child becomes fascinated with how the day always is light and the night is always dark, and continues to negotiating bedtimes. "Please, can I stay up a little longer?" "I don't want to go to bed." And this attention to curfews extends through the teenage years, negotiating once again, what time your teen must be home at night. Understanding about time is an important skill to have.

Children first learn about morning, noon, afternoon, and evening. Then comes an understanding of yesterday, today, and tomorrow. They learn to question "How long?" or "Are we there yet?" In today's world, they most likely see a digital clock first, before learning about analog clocks, because digital readouts are on ovens, computers, watches, and phones.

With this basic understanding of time, children can be introduced to the years, months, weeks, and days of a year, calendars, analog clocks, and distinguishing between AM and PM times. Once they understand the basics of an analog clock, they learn to distinguish between exact hours on the clock as compared to minute by minute time on a clock and second by second time. From this knowledge, it is only a small jump to estimating time needed to complete different tasks or activities.

Time Worksheet

1. What year is it now? _____

    What year was it last year? _____

    What year will it be next year? _____

    What year were you born? _____

2. What does chronological mean? _____

    _____

3. Draw a line from the months of the year to their chronological order in the year.

April	1
December	2
February	3
November	4
January	5
June	6
March	7
October	8
July	9
May	10
August	11
September	12

4. How many weeks are in a year? _____

5. What does PM mean?   Check the correct answer.

☐   Evening

☐   Noon

☐   Morning

6.   How many days are in a week?   _____

7.   Draw a line from the days of the week to their chronological order in the
     week.

Friday	1
Sunday	2
Monday	3
Saturday	4
Tuesday	5
Wednesday	6
Friday	7

8.   How many days are in a year?   _____

Look at the calendar below and answer the following questions:

January

SUNDAY	MONDAY	TUESDAY	WEDNESDAY	THURSDAY	FRIDAY	SATURDAY
					1	2
3	4	5	6	7	8	9
10	11	12	13	14	15	16
17	18	19	20	21	22	23
24 / 31	25	26	27	28	29	30

9.   On what day is the 25th of January? _____

10. What is the date of the second Monday in January? _____

11. How many Sundays are in January? _____

12. What day is New Year's Day? _____

13. What day is the last day of the month? _____

14. What day of the month is the 20th? _____

80

15. How many days are in the month of January? _____

16. What time is it on this clock?

_____

To what number is the hour hand pointing?

_____

To what number is the minute hand pointing?

_____

17. What is the time on this clock?

_____

18. What time is it on this clock?

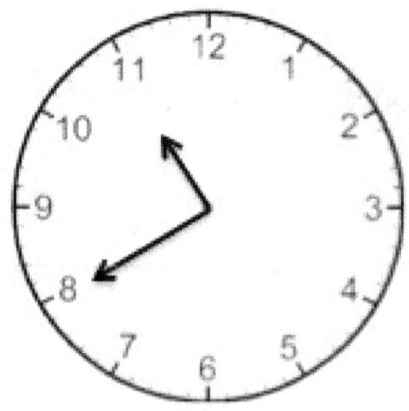

_____

19. What is the time on this clock?

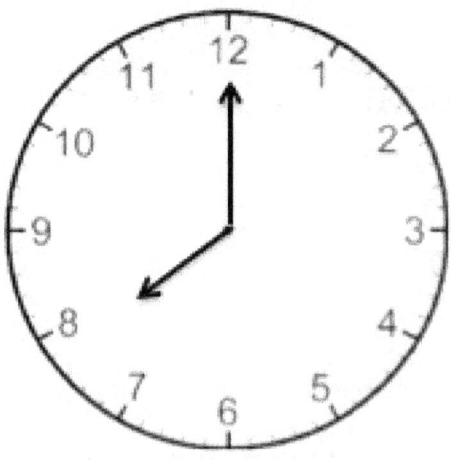

_____

20. What time is it on this clock?

_____

EXTRA CREDIT ACTIVITY

Make an analog clock with a paper plate to practice your time telling skills. The pattern is at the end of this lesson.

1) Get a paper plate. You can use either side of the plate.

2) Print out the patterns at the end of this lesson.

3) Cut out the patterns. One clock face and two hands.

4) Glue the clock face centered on the paper plate.

5) Punch a hole in the middle of the clock face. Punch holes at the end of the hands.

5) Thread the large hand and the small hand on the brass fastener and insert the fastener into the hole in the middle of the plate. Open the prongs of the fastener to secure the hands to the plate. You're done!

Using Shapes

As we learn more about shapes, we can start to use those shapes to increase spatial sense awareness. The 2D shapes that your child should know are square, rectangle, circle, triangle, hexagon, and pentagon. The 3D shapes that your child should know are cube, cuboid, sphere, cone, and pyramid. As we look more closely at these shapes, we will see how they affect our lives and our mathematical subconscious perceptions.

Using Shapes Worksheet

In the graphic below, color the shapes as indicated (Note: If you are doing this on an eReader, then you will first need to draw the shapes!):

1.   Color all of the squares red.

2.   Color all of the rectangles blue.

3.   Color all of the circles yellow.

4.   Color all of the triangles in green.

5.   Choose items from the list below that can be made with square shapes. Check all that apply.

☐ sphere	☐ cube	☐ checker board
☐ cow	☐ house	☐ quilt
☐ tree	☐ fish	☐ chair

☐ bucket   ☐ car   ☐ airplane

6.   Name four items that can be made by using rectangle shapes.

     _____   _____

     _____   _____

7.   Look in a magazine and find four objects that have circle shapes. Paste them on the back of this paper.

8.   Cut out six identical squares. Using tape, assemble them into a cube.

9.   Cut out six shapes: Four identical rectangles and 2 identical squares. Using tape, assemble them into a cuboid.

10.   Cut out a circle. Fold it in half and then cut along the fold to make two halves. Take one half circle and fold the straight edge in half. Bring the two edges of the half circle together to form a cone. Tape the edges together.

11.   Cut out four identical triangles. Measure the base of the triangles. Cut out a square with sides that are the same length as the base of the triangles. Using tape, assemble them into a four-sided pyramid.

12.   Look in a magazine and find six objects that have triangle shapes. Paste them on the back of this paper.

13.   Make a triangle fish. Cut out two triangles that are the same size from two different colors of paper. Paste one triangle on top of the point of the other triangle. Using a marker, add an eye and a mouth. Decorate the fish with glitter and markers. Mount your

finished fish on a sheet of blue paper and draw waves for the fish to swim in and bubbles coming up from his mouth.

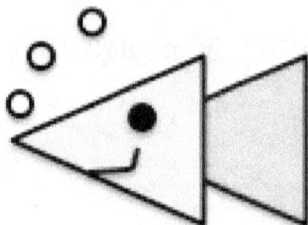

14. Make a bull's eye with circles. Cut out five sizes of circles, each in a different color of paper. Paste them on top of each other in order of size with the largest circle on the bottom the smallest circle on top. Make sure each circle is centered on top of the next circle. Mount your bull's eye on a sheet of colorful construction paper and hang it on your wall.

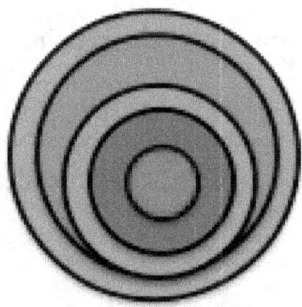

15. Make a snowman from hexagons. Cut out three different sizes of hexagons in white construction paper. Paste the three hexagons stacked on top of each other with the largest one on the bottom and the smallest one on the top. Add a hat in the shape of a triangle and two little round eyes. Paste on a heart shaped mouth. Add three round buttons glued on the second hexagon along with two straight line arms. Color some background for your snowman to enjoy! Pattern pieces are on the last page of this lesson.

16. What is the most famous pentagon

building in the world?

_____

17.   What shape makes up the design on a soccer ball?

        _____

18.   Draw a dinosaur and put scales on his back using pentagons. You can color them or cut
        them out with construction paper and glue them onto the back of the dinosaur. Use this
        dinosaur as a guide. You can even make baby dinosaurs!

19.   Name the four shapes in this graphic:

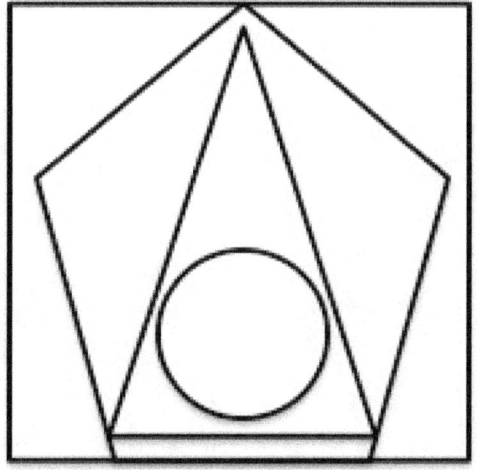

_____

_____

_____

_____

20.   How many sides does a hexagon have?

_____

How many sides does a pentagon have?

_____

Weight

Weight is something that we feel when we sit in a chair or climb up a flight of stairs. Kids experience the effect of weight every time they go to an amusement park. Gravity, on the other hand, even though it is working all of the time, we don't notice it. When we are on the amusements rides, we might, on occasion feel weightlessness for a few seconds.

Mass and weight are two different things. Weight is how heavy something is. Mass is how much space something occupies. Matter makes up mass and can be heavy or light. If you take a brick and a sponge that is the exactly same size as the brick, the brick will be heavier. The mass of the brick and the sponge is the same because they are the same size and take up the same amount of space. The matter that makes up the contents of the brick is heavy while the mass that makes up the sponge is light.

Weight Worksheet

Figure out what should be in the blanks for questions 1 through 3:

1. _____ is how heavy something is.

2. _____ is how much space something occupies.

3. _____ makes up mass and can be heavy or light.

4. When you make a guess or prediction, it is called

    ☐   an observation.

    ☐   an hypotheses.

    ☐   a question.

5.   If you want to weigh heavy objects, you need what type of scale?

☐   kitchen scale

☐   balance scale

☐   bathroom scale

6.   If you want to weigh light weight items, you need what type of scale?

☐   balance scale

☐   bathroom scale

☐   kitchen scale

7.   Step on the bathroom scale. How much do you weigh? _____

8.   Using the balance scale, which item was heavier?

☐   marble

☐   cotton ball

☐   Styrofoam ball

9.   When you are weighing two items on the balance scale and the beam of the scale does not dip, what does that mean?

_____

_____

10.   What the object that is heavier?

11.   What the object that is lighter.

12.   What the object that is heavier.

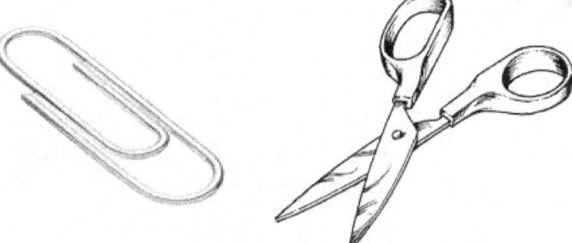

13.   What the object that is lighter:

14. What the object that is heaviest:

15. What the object that is lightest:

 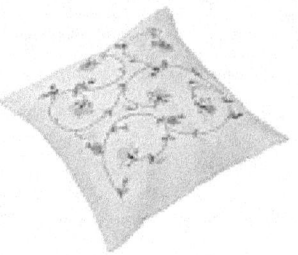 

16. Number the items in order from lightest to heaviest. 1 is for lightest, 2 is the middle weight, and 3 is for the heaviest.

17. Number the items in order from heaviest to lightest. 1 is for heaviest, 2 is the middle weight, and 3 is for the lightest.

18. Number the items in order from lightest to heaviest. 1 is for lightest, 2 is the middle weight, and 3 is for the heaviest.

  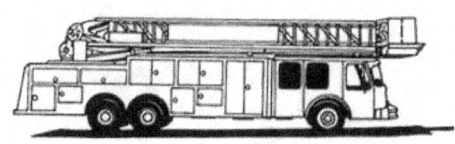

19. Number the items in order from heaviest to lightest. 1 is for heaviest, 2 is the middle weight, and 3 is for the lightest.

20.   What is the balancing scale.

Workbook Answer Key

Algebra Worksheet Answer Key

6	13	84	4	33	91	2
+4	+6	+11	+1	+12	+7	+0
10	19	95	5	45	98	2

$4 + \underline{3} = 7$ $\underline{3} + 6 = 9$ $55 + \underline{6} = 61$ $24 + \underline{9} = 33$

$5 + 5 = \underline{10}$ $8 + 7 = \underline{15}$ $15 + 18 = \underline{33}$ $22 + 32 = \underline{54}$

Solve for X:

$5 + X = 10$ $12 + X = 24$ $X + 7 = 11$ $45 + X = 55$
   5   **12**   **4**   **10**

$4 + X = 21$ $9 + X = 32$ $X + 5 = 15$ $47 + X = 74$
   17   **24**   **10**   **27**

3	8	5	4	6	9	7
X4	X1	X3	X6	X3	X7	X0
12	8	15	24	18	63	0

5 X 7 = <u>35</u>   8 X 2 = <u>16</u>   12 X 12 = <u>144</u>   9 X 8 = <u>72</u>

Solve for X:

$5 X = 10$ $7 X = 21$ $9 X = 81$ $11 X = 22$

2   **3**   **9**   **2**

$9 X = 9$ $10 X = 50$ $8 X = 24$ $6 X = 0$

1   **5**   **4**   **0**

Challenge Questions: $\dfrac{32}{X} = 4$ $\dfrac{56}{X} = 8$

8   **7**

Capacity Worksheet Answer Key

Definitions: (Write answers on separate piece of paper.)

5. Who is Archimedes? <u>GREEK MATHEMATICIAN WHO INVENTED MANY OF THE MATHEMATICAL PROCESSES WE USE TODAY</u>

6. What did Archimedes invent? <u>USING WATER DISPLACEMENT TO MEASURE VOLUME</u>

7. What does displaced mean? <u>DIFFERENCE OF THE INITIAL POSITION OF WATER WHEN AN OBJECT IS ADDED TO IT</u>.

8. What does estimate mean? <u>APPROXIMATE CALCULATION</u>

9. What does volume mean? <u>TOTAL AMOUNT OF SOMETHING</u>

10. What does capacity mean? <u>THE AMOUNT THAT CAN BE HELD IN A CONTAINER</u>.

11. Which has the largest capacity? Circle the correct item.

<u>O</u>

12.	Which takes up the smallest volume? Circle the correct item.

Ο

13.	Which has the smallest capacity? Circle the correct item.

Ο

14.	Which container will hold the most jelly beans?

Ο

15.	Which shown container will hold the most water?

Ο

16.	Which shown container will hold the least amount of air?

Write answers to questions below on separate piece of paper.

17. When you estimate the number of jelly beans in a container, how do you determine the exact amount of jelly beans that are actually in the container? COUNT THEM.

18. Make up your own problem to explain capacity and show your parent or teacher how to do it.

19. Using a measuring cup, pour ¼ cup of water into a cup. Add ¾ cup of water to that amount. How much water is now in the cup? 1 cup

20. Take a cup of rice. Remove ¼ cup of rice from the cup of rice. How much rice is left in the cup? ¾ cup

Fractions Worksheet Answer Key

1. Write the parts of these circles in fractions:

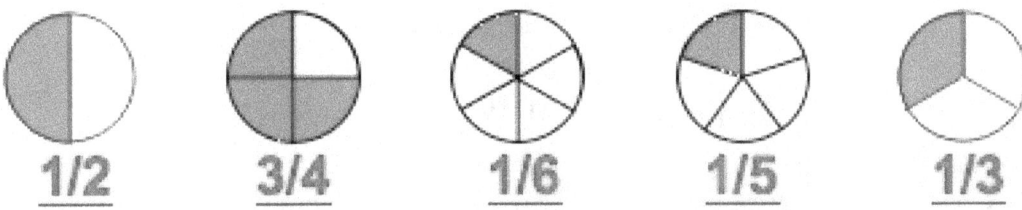

2. Divide the circles below into the fractions as shown: (Color in the numerators)

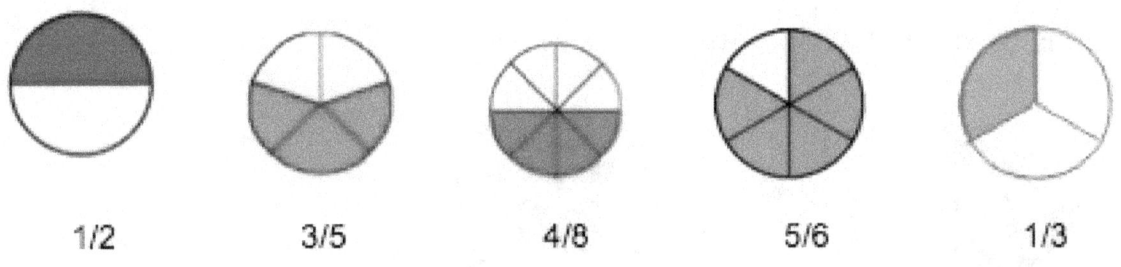

3. Color in the numerators as indicated by the fractions below:

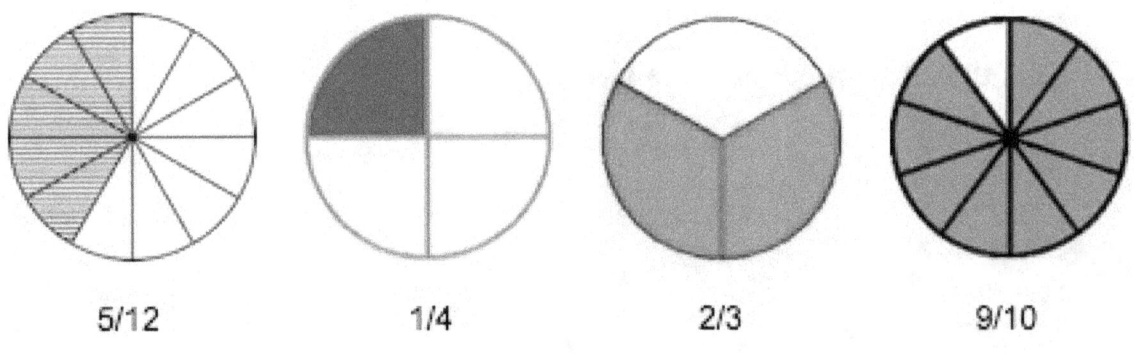

4. What is a numerator?   <u>THE TOTAL NUMBER OF THE PARTS CHOSEN</u>

5. What is a denominator? <u>THE TOTAL NUMBER OF THE PARTS</u>

6. What is a fraction? <u>PART OF A WHOLE</u>

7. In the fraction 1/5, what is the numerator? <u>1</u>

8. In the fraction 12/13, what is the denominator? <u>13</u>

9. What is the number that represents 4/4? <u>1</u>

10. In the fraction 12/12, what is the numerator? <u>12</u>

11. In the fraction 9/10, what is the denominator? <u>10</u>

12. What fraction describes these circles?

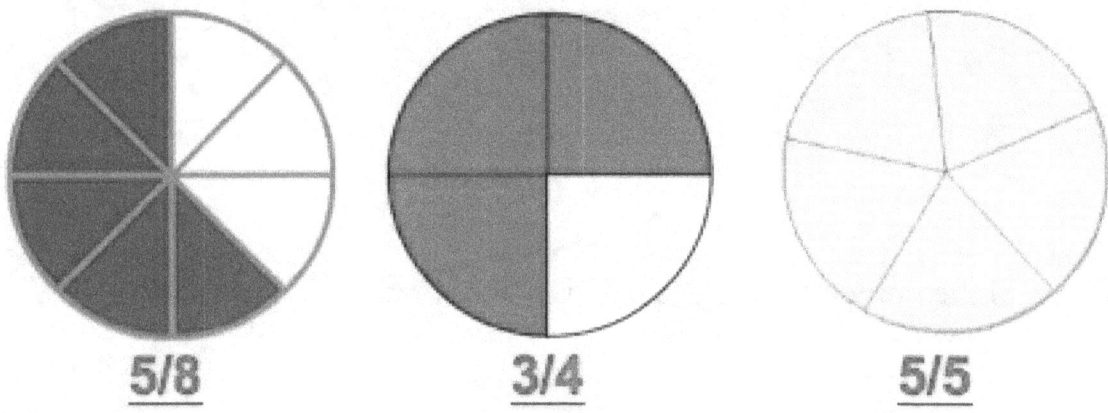

<u>5/8</u>   <u>3/4</u>   <u>5/5</u>

13. What is the numerator for these circles?

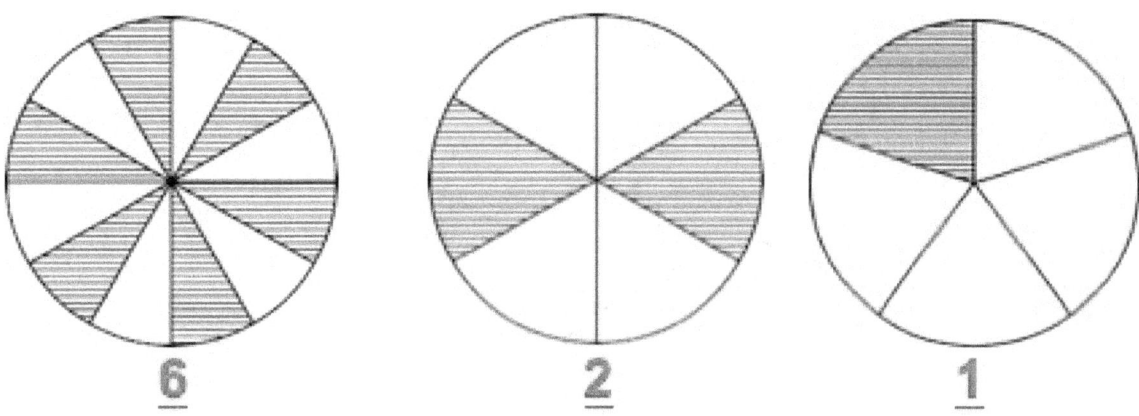

14. What is the denominator for these circles?

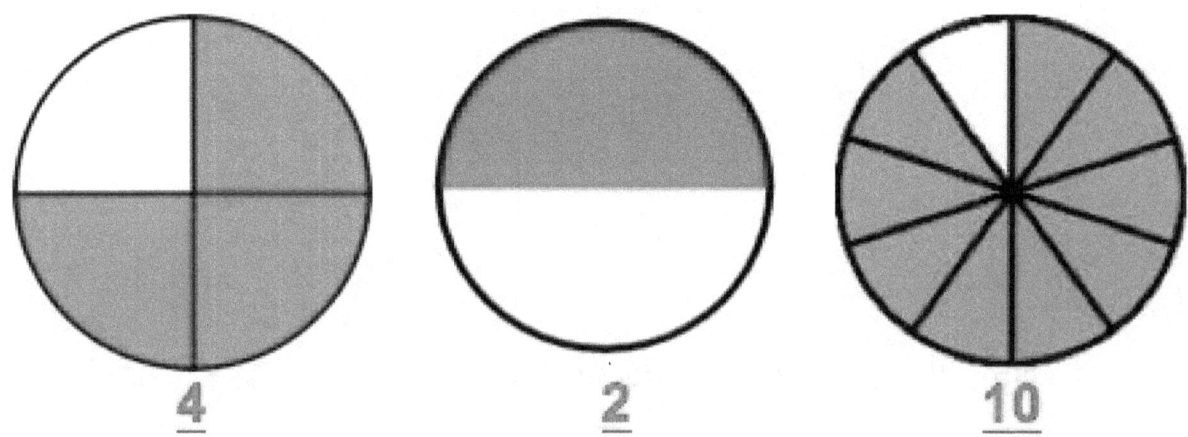

Answer the following questions with the correct fraction:

15. If you have a pizza with 6 slices and you eat 1 slice, you have 5/6 of the pizza left over.

16. If you have a cookie and you eat half of it, you have 1/2 of it left over.

17. If a pie has six slices and you eat two of the slices, you have 4/6 of the pie left over.

18. Draw a circle that represents the fraction 1/2.

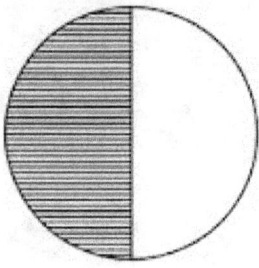

19. Draw a circle that represents the fraction 3/3.

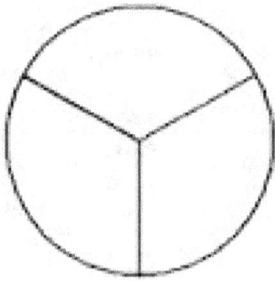

20. Draw a circle that represents the fraction ¼.

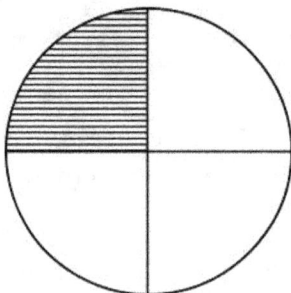

Length Worksheet Answer Key

5.   Looking at your ruler, how many inches are in a foot?   <u>12</u>

6.   Mark on the ruler below a length of 2 and 1/4 inches:

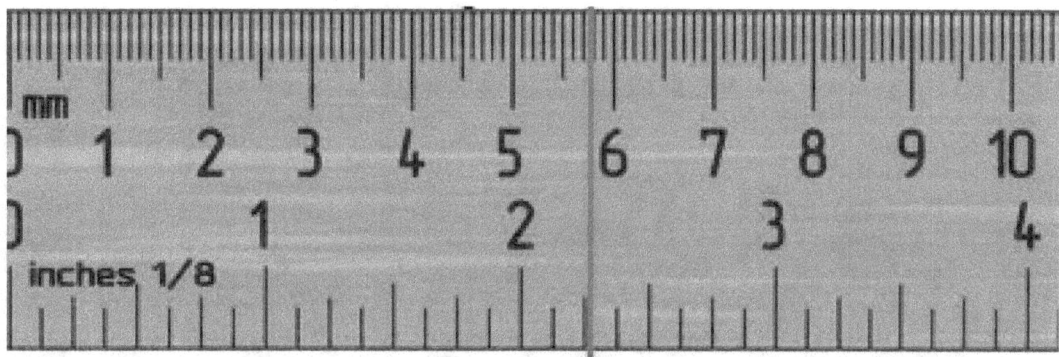

7.   Mark on the ruler below a length of 7 centimeters:

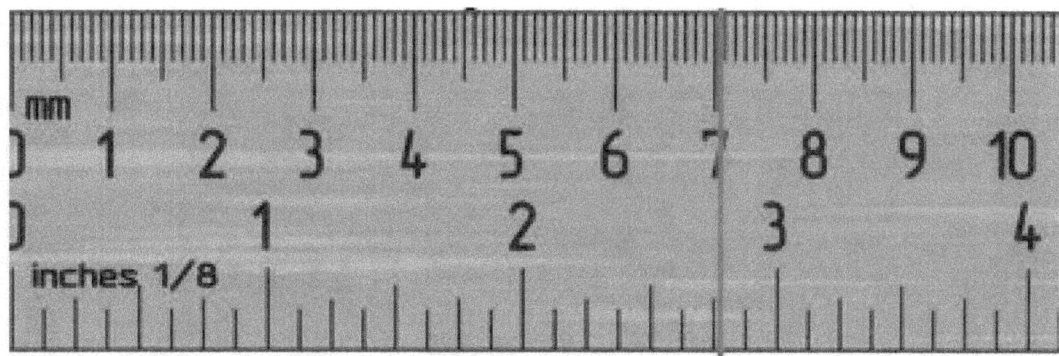

8.   Mark on the ruler below a length of 3 and 1/2 inches:

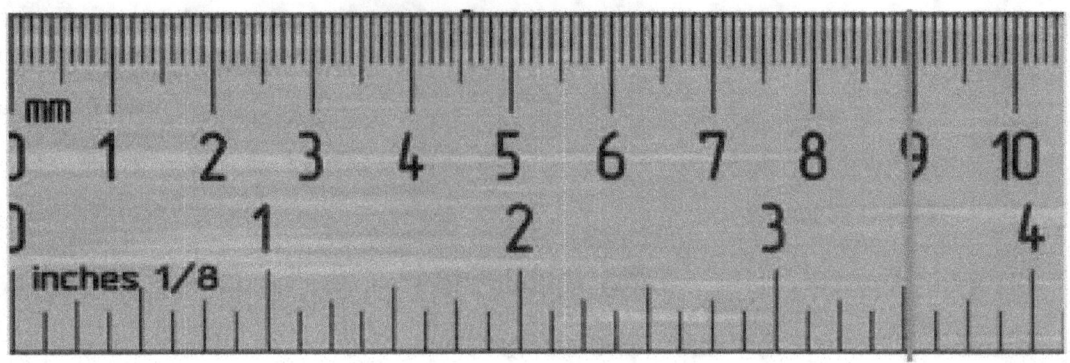

9. Mark on the ruler below a length of 4 and 1/2 centimeters:

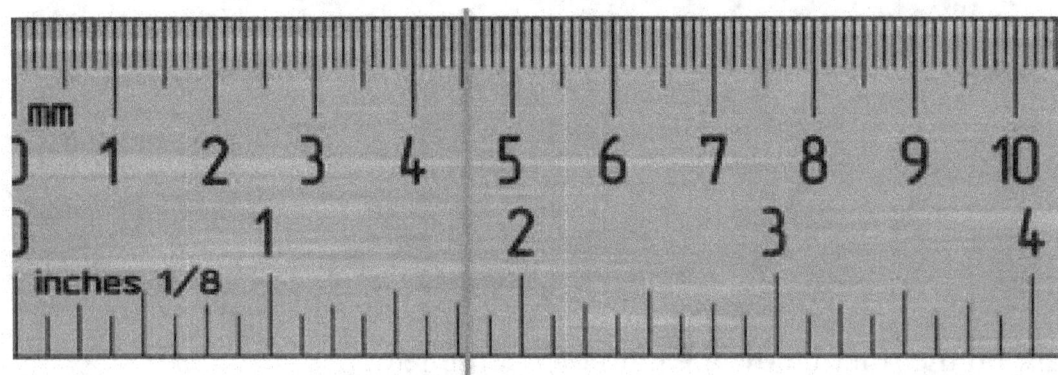

10. Mark on the ruler below a length of 3 and 3/4 inches:

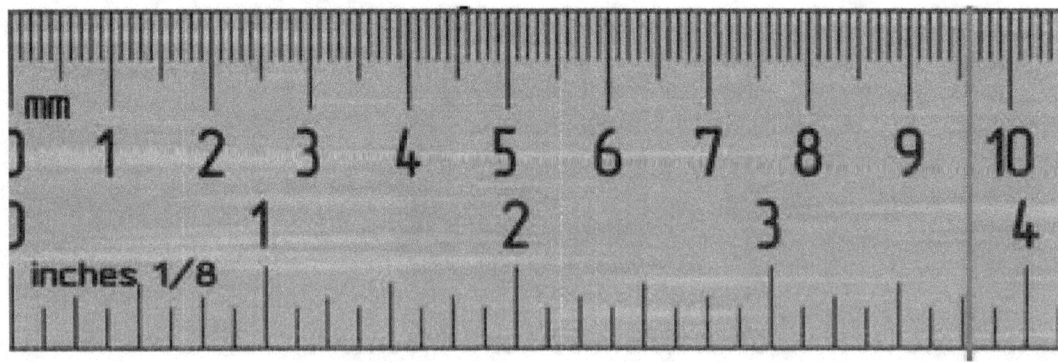

11. Looking at the ruler above, approximately how many centimeters are in 2 inches? <u>5</u>

12. Looking at the ruler above, approximately how many centimeters are in 4 inches? _10_

13. What is a definition of length? _____

MEASURES HOW LONG SOMETHING IS.

14. Describe two ways you can measure things:

USING AN ITEM LIKE YOUR FEET TO MEASURE SOMETHING
OR

USING A RULER OR TAPE MEASURE.

15. Briefly explain the problem the King needed to solve in the book "How Big Is a Foot":

HOW TO FIGURE OUT WHAT SIZE BED THE QUEEN NEEDED.

16. Find an item that is 1 inch in length.

What item did you measure that was 1" in length? _____

17. Find an item that is 1 foot in length.

What item did you measure that was 1' in length? _____

18. Looking at your ruler, how many 1/4 inch lengths are in 1 inch?

    <u>4</u>

19. Looking at your ruler, how many 1/2 inch lengths are in 1 inch?

    <u>2</u>

20. Looking at your ruler, how many 1/8 inch lengths are in 1 inch?

    <u>8</u>

Money Worksheet Answer Key

Look at the coins in the middle column. Count up the value of the coins. Write the number value of the coins in the right hand column. The first line is an example.

Ex	(quarter)	25¢
1	(nickel) + (nickel) + (nickel) + (nickel)	20¢
2	(dime) + (dime) + (dime) + (dime) + (dime)	50¢
3	(penny) + (penny) + (penny) + (penny) + (penny)	5¢
4	(quarter) + (quarter)	50¢
5	(quarter) + (nickel) + (nickel) + (penny)	36¢
6	(penny) + (nickel) + (dime) + (quarter)	41¢
7	(penny) + (penny) + (penny) + (dime) + (dime) + (dime)	33¢
8	(nickel) + (nickel) + (nickel) + (penny)	16¢

9	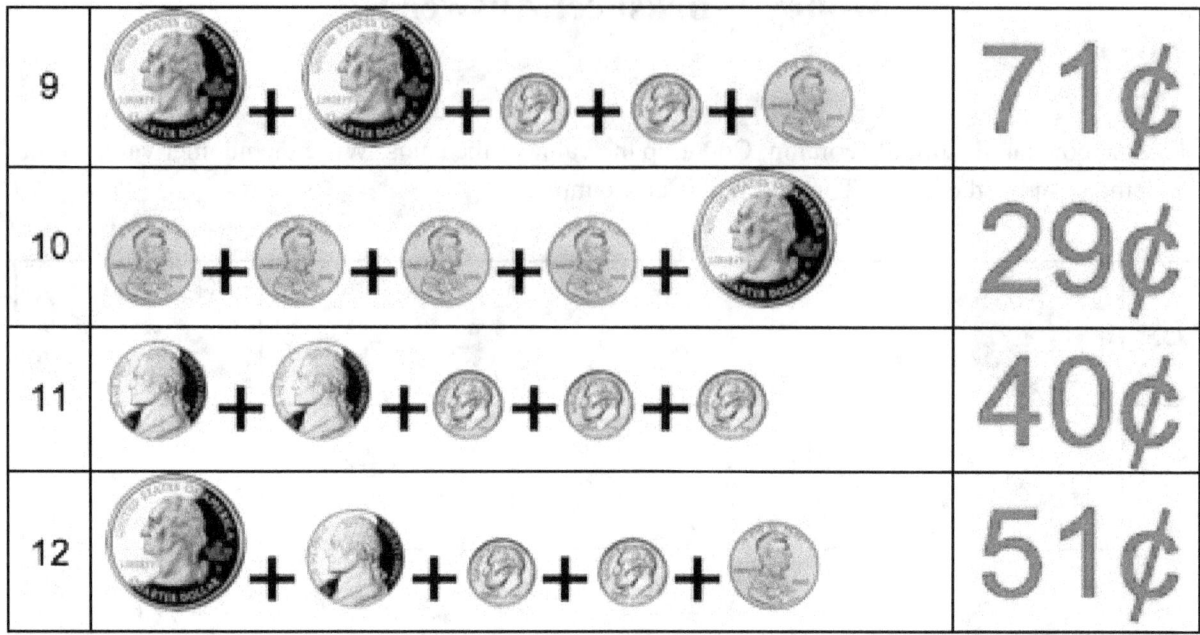	71¢
10		29¢
11		40¢
12		51¢

Look at the paper currency in the middle column. Count up the value of the paper currency. Write the number value of the paper currency in the right hand column. The first line is an example.

EX	$1 bill	$5 bill	**$6.00**
13	$10 bill	$20 bill	$30.00
14	$1 bill	$10 bill	$11.00
15	$5 bill	$20 bill	$25.00
16	$20 bill	$1 bill	$21.00

17	$5 bill	$10 bill	$15.00
18	$1 bill	$1 bill	$2.00
19	nickel + dime + penny + $20 bill		$20.16
20	$5 bill + quarter + quarter		$5.50

Number Sense Worksheet Answer Key

1. Starting in the upper left hand corner and going across one row at a time from top to bottom, number each of the squares from 1 to 100.

1	2	3	4	5	6	7	8	9	10
11	12	13	14	15	16	17	18	19	20
21	22	23	24	25	26	27	28	29	30
31	32	33	34	35	36	37	38	39	40
41	42	43	44	45	46	47	48	49	50
51	52	53	54	55	56	57	58	59	60
61	62	63	64	65	66	67	68	69	70
71	72	73	74	75	76	77	78	79	80
81	82	83	84	85	86	87	88	89	90
91	92	93	94	95	96	97	98	99	100

2. What is a tip for remembering symbols for more than and less than?

    The opening of the sideways "V" is open BIG and is always next to the bigger side; the point of the sideways "V" is small and always points to the smaller side.

3. What are the three comparison symbols and what do they mean?

    > means MORE THAN

< means LESS THAN

= means EQUAL TO

4.   Write the correct comparison symbol in the blank boxes for each set of numbers.
     (< = >)

1)	82	>	5
2)	37	=	37
3)	1	<	99
4)	55	>	21
5)	29	<	30
6)	99	=	99
7)	16	>	10
8)	44	<	49
9)	68	<	70
10)	71	>	17

5.   Starting in the upper left hand corner and going across one row at a time from top to bottom, number each of the squares counting by 2's.

2	4	6	8	10	12	14	16	18	20
22	24	26	28	30	32	34	36	38	40
42	44	46	48	50	52	54	56	58	60
62	64	66	68	70	72	74	76	78	80
82	84	86	88	90	92	94	96	98	100

6.   Starting in the upper left hand corner and going across one row at a time from top to bottom, number each of the squares counting by 5's from 5 to 100.

5	10	15	20	25	30	35	40	45	50
55	60	65	70	75	80	85	90	95	100

7.   Starting in the upper left hand corner and going across one row at a time from top to bottom, number each of the squares counting by 10's.

10	20	30	40	50	60	70	80	90	100

8.   Add the numbers for the sums to these problems:

$$\begin{array}{cccccccc}
1 & \quad 2 & \quad 13 & \quad 14 & \quad 15 & \quad 16 & \quad 7 \\
\underline{8} & \quad \underline{9} & \quad \underline{10} & \quad \underline{11} & \quad \underline{12} & \quad \underline{13} & \quad \underline{1} \\
9 & \quad 11 & \quad 23 & \quad 25 & \quad 27 & \quad 29 & \quad 8
\end{array}$$

9.   Subtract the numbers to answer these problems:

$$\begin{array}{ccccccc}
20 & \quad 19 & \quad 18 & \quad 17 & \quad 16 & \quad 15 & \quad 14 \\
\underline{1} & \quad \underline{0} & \quad \underline{10} & \quad \underline{3} & \quad \underline{11} & \quad \underline{10} & \quad \underline{4} \\
21 & \quad 19 & \quad 28 & \quad 20 & \quad 27 & \quad 25 & \quad 18
\end{array}$$

10.   What is the inverse of this addition problem $11 + 6 = 17$?

$$6 + 11 = 17 \qquad 17 - 6 = 11 \qquad 17 - 11 = 6$$

11.   What is the inverse of this subtraction problem $20 - 11 = 9$?

$$20 - 9 = 11 \qquad 11 + 9 = 20 \qquad 9 + 11 = 20$$

12.   What is the inverse of this addition problem $10 + 17 = 27$?

$$17 + 10 = 27 \qquad 27 - 17 = 10 \qquad 27 - 10 = 17$$

13.   What is the inverse of this subtraction problem $11 - 3 = 8$?

$$11 - 3 = 8 \qquad 3 + 8 = 11 \qquad 8 + 3 = 11$$

14.   What number is ten less than 34?   <u>24</u>

15.   What number is ten more than 74?   <u>84</u>

16.   What number is ten less than 99?   <u>89</u>

17.   What number is ten more than 11?   <u>21</u>

18.   What number is one more than 37?   <u>38</u>

19.   What number is one less than 83?   <u>82</u>

20.   What number is one less than 46?   <u>45</u>

Operations Worksheet Answer Key

Using the commutative property of addition, rewrite these problems in a different way.

1. $2 + 8 + 10 = 20$ $2 + 8 = 10$ so $10 + 10 = 20$

2. $5 + 5 + 20 = 30$ $5 + 5 = 10$ so $10 + 20 = 30$

3. $2 + 2 + 7 = 11$ $2 + 2 = 4$ so $4 + 7 = 11$

Using the associative property of addition, rewrite these problems in a different way.

4. $5 + 7 = 12$ $7 + 5 = 12$

5. $8 + 1 = 9$ $1 + 8 = 9$

6, $6 + 4 = 10$ $4 + 6 = 10$

7. What is the meaning of the equal sign?   Same or equivalent

_____

Look at the problems below with an equal sign. Decide if they are true or false. Circle your answer.

8. $8 + 1 = 9$ TRUE   FALSE

9. $2 + 1 + 1 + 4$ TRUE   FALSE

10. $5 = 7$ TRUE   FALSE

11. $3 + 5 = 8$ TRUE   FALSE

12. $6 + 2 = 4 + 5$ TRUE   FALSE

13. $4 + 4 = 7$ TRUE   FALSE

14. $1 + 1 + 1 = 3$ TRUE   FALSE

15. $7 + 3 = 10 - 1$ TRUE   FALSE

16. Write the numbers 10 to 100, counting by tens. _____

    10   20   30   40   50   60   70   80   90   100

17. Write the ODD numbers only from 1 to 20: _____

    1   3   5   7   9   11   13   15   17   19

18. Write the EVEN numbers only from 1 to 20: _____

    2   4   6   8   10   12   14   16   18   20

19. What is the definition of sum?   TOTAL AMOUNT

Give an example of a sum: $5 + 5 = 10$ 10 IS THE SUM OF 5 + 5

20. Will the sum of 3 + 3 be an even or odd number?   EVEN

Will the sum of 3 + 4 be an even or odd number?   ODD

Will the sum of 11 + 9 be an even or odd number?   EVEN

Will the sum of 7 + 10 be an even or odd number?   ODD

Patterns Worksheet Answer Key

1. Describing Patterns: Carefully look at the two diagrams below and answer the questions:

Diagram A

Diagram B

☐ How are the two patterns alike? _____

THE THIRD AND NINTH BOXES ARE PURPLE. THE DIAGRAMS

HAVE THE SAME NUMBER OF BOXES. THE SECOND, FOURTH,

EIGHTH, AND TENTH BOXES ARE WHITE.

_____

☐ How are the two patterns different? _____

DIAGRAM A ALTERNATES WITH ONE PURPLE BOX AND THEN

ONE WHITE BOX. DIAGRAM B ALTERNATES WITH TWO WHITE

BOXES AND ONE PURPLE BOX.

_____

2.   What is a 5-unit sequence? _____

A PATTERN WITH 5 UNITS OR NUMBERS IN IT.

3.   What is a repeating pattern? _____

A SEQUENCE OCCURING MORE THAN ONE TIME.

4.   Repeating Patterns: Using markers that are the same colors as the 4-unit
     pattern, color in the remaining boxes, following the pattern exactly until you
     run out of boxes.

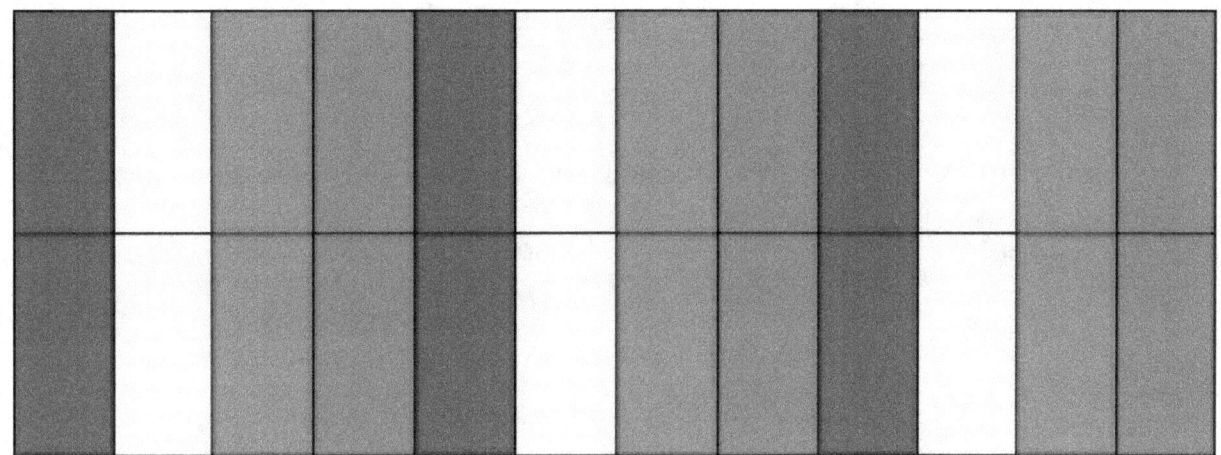

5. Answer the following questions about the Repeating Patterns above:

☐ How many boxes are in the above unit pattern? <u>4</u>

☐ How many times did you repeat your pattern? <u>6</u>

☐ Were any column patterns created? <u>YES</u>

Describe them: <u>BY REPEATING THEM, THERE WERE 12</u>

<u>COLUMNS IN FOUR COLORS</u>.

☐ Were any diagonal patterns created? <u>NO</u>

Describe them: _____

6. Using the following number sequence, repeat this 5-unit pattern sequence 2 more times, making 3 sequences.

13579 13579 13579

7. Look at the growing pattern below and add the next sequence:

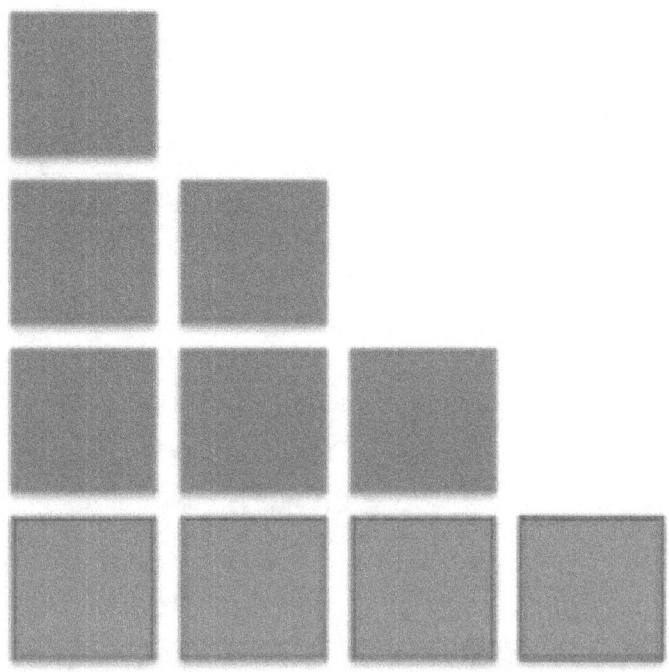

8. Look at this growing sequence of circles and add the next sequence:

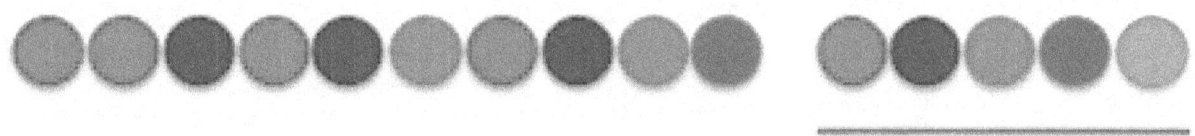

9. Look at this growing sequence of numbers and add the next sequence:

9 9 7 9 7 5 9 7 5 3

9 9 7 9 7 5 9 7 5 3 9 7 5 3 4

Are the patterns below repeating or growing?

10. <u>REPEATING</u>

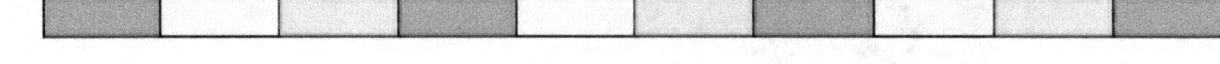

11. <u>GROWING</u>

12. <u>GROWING</u>

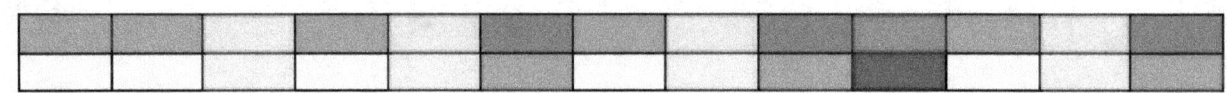

13. <u>REPEATING</u>

123A123A123A23A

14. <u>GROWING</u>

1 1 A 1 A 2 1 A 2 B

15. <u>GROWING</u> _

16. What shapes complete the last sequence?

17. What numbers complete the last sequence?

1 1 2 1 2 3 1 2 3 4 1 2 3 4 5 1 2 3 4 5 6   1 2 3 4 5 6 7

18. Create your own repeating pattern using at least 3 sequences:

1 2 3 4 5 1 2 3 4 5 1 2 3 4 5 1 2 3 4 5

19. Create your own growing pattern using numbers and 4 sequences:

1 1 2 1 2 3 1 2 3 4

20. Create your own growing pattern using shapes in 3 sequences:

9 9 8 9 8 7

Positions Worksheet Answer Key

1. Which box is above the purple box?   PINK

2. Is the green box above or below the purple box?   BELOW

3. Which box is in the middle?   PURPLE

4.

    Circle the best answer:

        Left   Right

7. In what position is the little man with the red hair and gray suit?

    Circle the correct answer:

    Left   Center   **Right**

8.   In what position is the girl with the

     red and white uniform?

     Circle the correct answer:

     Right   **Left**

11. Is the yellow star in the middle, on the top

     or on the bottom? TOP

     _____

     What is the position of the red star? CENTER

     _____

12. What does position mean? _____

     WHERE SOMETHING IS LOCATED

     _____

13. What is on the top of this ice cream?

CHERRY

What is on the bottom?

BOWL

14. Is the square inside, on, or outside the circle?

ON THE CIRCLE

15. Is the heart inside, on, or outside the square?

INSIDE THE SQUARE

16. Is the star inside, on, or outside the Triangle?

OUTSIDE THE TRIANGLE

Locate the position of the following objects on a 3 X 3 grid:

17. Circle the object in the top right position.

    PINK HEART

18. Draw a square around the object in the

    center position.

    YELLOW SUN

19. Draw an "X" over the object in the bottom

    left position.

    ORANGE STAR

20. Draw a triangle over the object in the

    top center position.

    RED CIRCLE

Shapes Worksheet Answer Key

1. Circle the shapes that are two dimensional (2D) below:

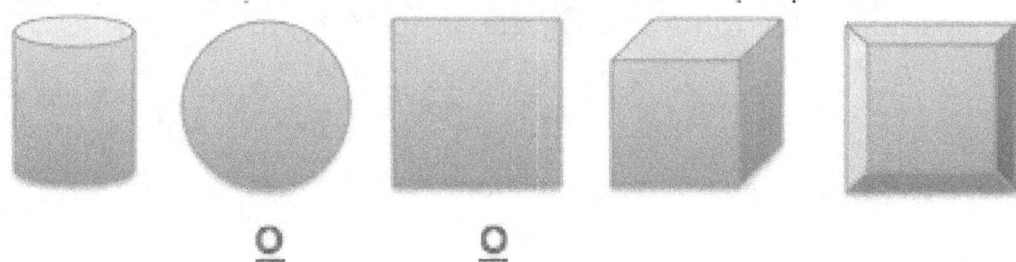

2. Circle the shapes that are three dimensional (3D) below:

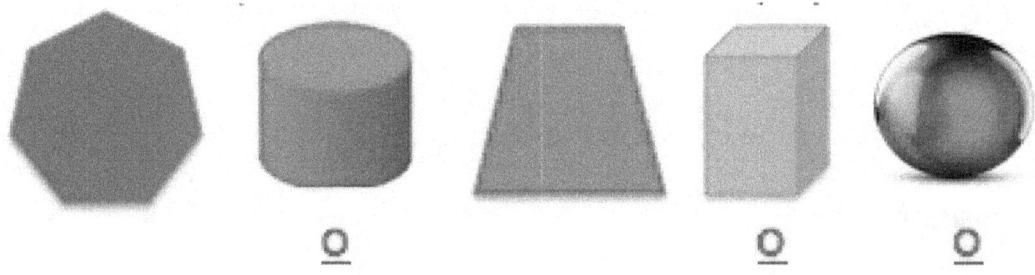

3. What makes an object 2D (two dimensional)? _____

    IT IS FLAT WITH ONLY LENGTH AND WIDTH.

4. What makes an object 3D (three dimensional)? _____

    IT HAS VOLUME WITH LENGTH, WIDTH, AND DEPTH.

5. Circle the closed shapes below:

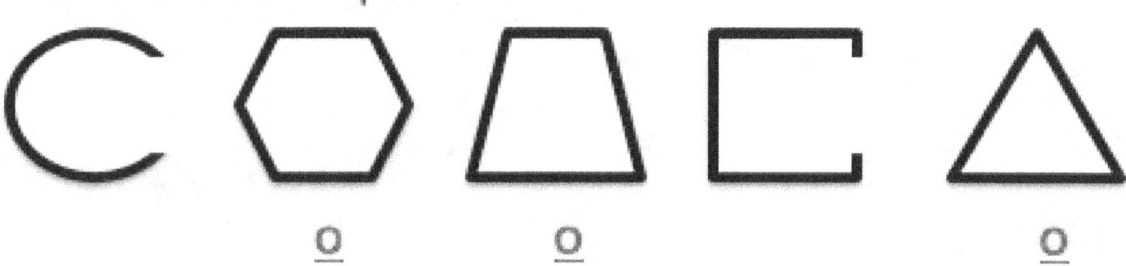

6. Circle the open shapes below:

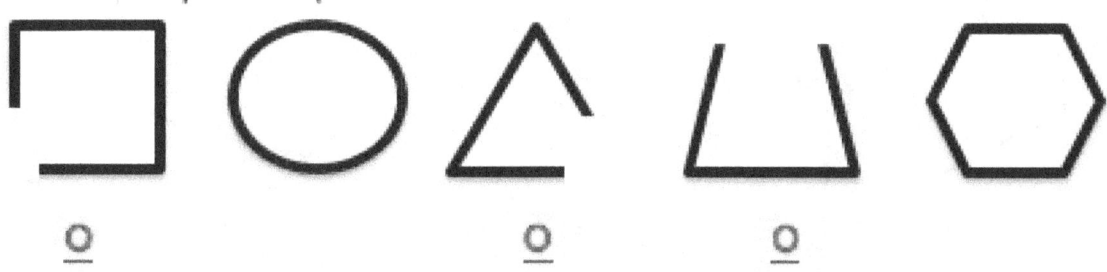

7. Circle similar shapes:

3 CIRCLES

8. Circle dissimilar shapes:

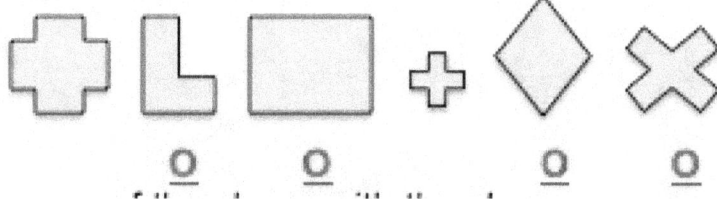

9. Draw a line connecting the name of the shape with the shape:

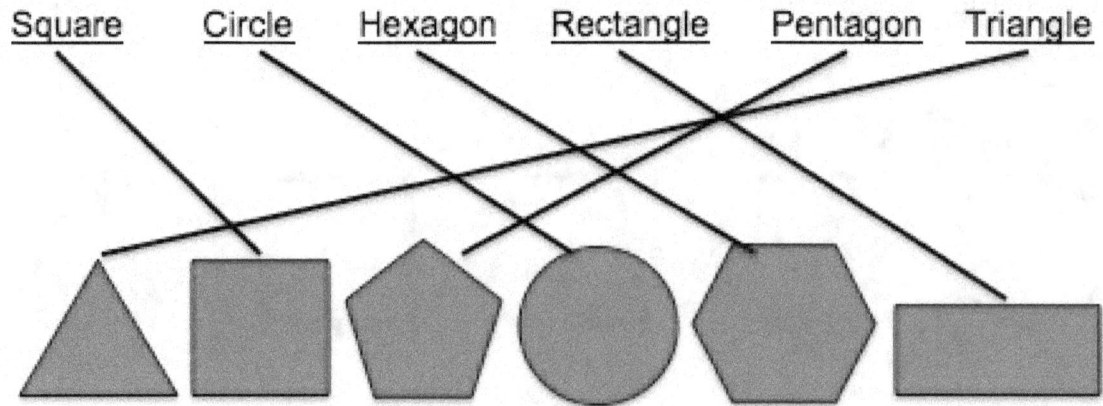

10. Draw a line connecting the name of the shape with the shape:

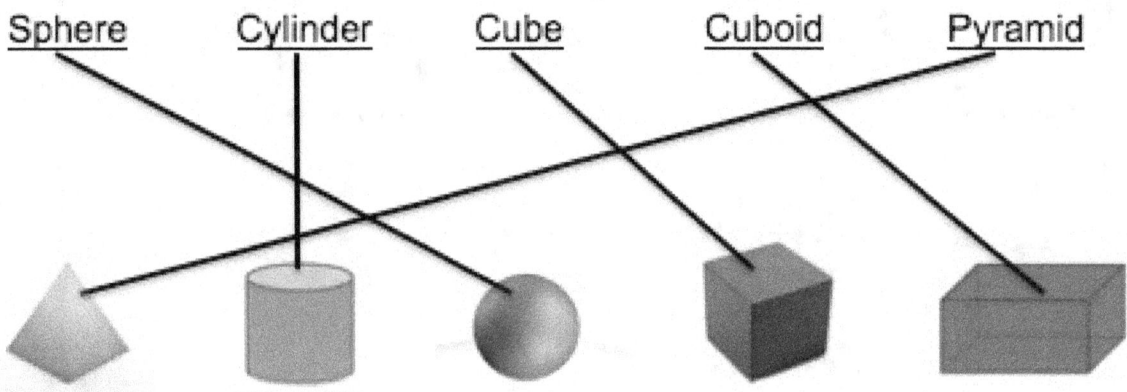

11. Draw a line connecting the name of the shape to its comparison:

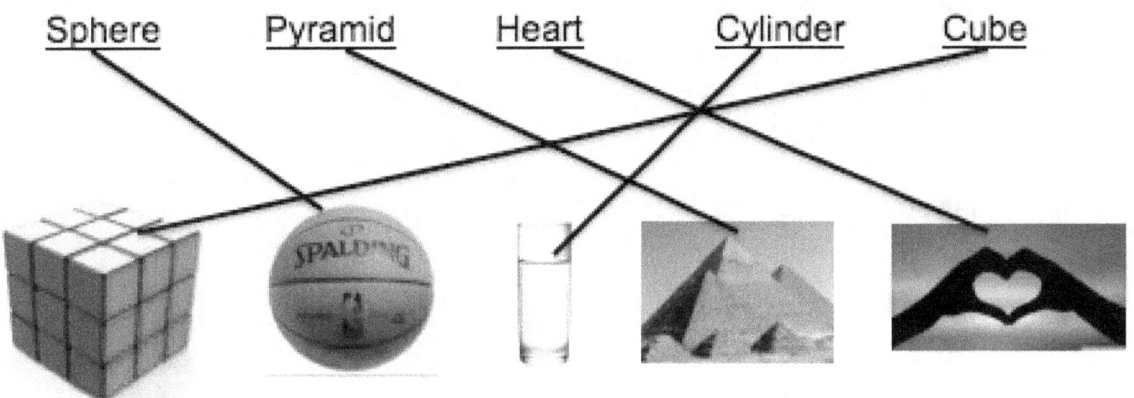

Sphere   Pyramid   Heart   Cylinder   Cube

12. What shape looks like this? ✓ the correct answer.

☐ square

☐ **pentagon**

13. What shape looks like this? ✓ the correct answer.

☐ **cuboid**

☐ pyramid

14. What shape looks like this? ✓ the correct answer.

☐ rectangle

☐ **cylinder**

15. Which planar shape relates to this figure?

☐ **rectangle**

☐ circle

☐ triangle

15. Which planar shape relates to this figure?

☐ pentagon

☐ **square**

☐ heart

16. Which planar shape relates to these objects?

☐ triangle

☐ **hexagon**

☐ square

17. Is this object symmetrical?

☐ yes

☐ **no**

18. Is this animal symmetrical?

☐ **yes**

☐ no

19. Circle the objects that have symmetry:

o   o

20. Draw a line through the object below to show symmetrical halves:

138

Spatial Sense Worksheet Answer Key

2.   How many sides does it have?

       <u>6</u>

3.   What 2D shape makes a cube?

       <u>SQUARE</u>

Look at a cylinder shape.

4.   If you cut the cylinder and lie it flat, what 2D shape do you have?

       <u>RECTANGLE</u>

5.   If you want to put a top and bottom on the cylinder, what shape will you need to do that?

       <u>CIRCLE</u>

6.   Look at this pyramid. How many <u>triangles</u> make up this pyramid?

     <u>4</u>

7.   What 2D shape do you need to make a 3D shaped cone?

     ☐   circle

     ☐   square

     ☐   half circle

8.   In the cone shape above, what shape is the base of the cone?

     <u>CIRCLE</u>

Look at this cuboid shape.

9.   How many sides does it have?

     <u>6</u>

10. What 2D shapes make this cuboid?

     <u>RECTANGLE</u> and <u>SQUARE</u>

8.   How many squares does it take to make a cube?

☐   3

☐   4

☐   6

9.   How many sides does a <u>closed</u> cone shape have?

☐   3

☐   2

☐   1

10.   How many sides does a <u>closed</u> cylinder shape have?

☐   1

☐   2

☐   3

11.   Explain how a box is made from squares. _____

<u>TAPE SIX SQUARES TOGETHER TO MAKE A CUBE.</u>

_____

_____

12.   Explain how to make a cone from paper. _____

TAKE A HALF CIRCLE, FOLD IT ALNG THE STRAIGHT EDGE,

DIVIDING IT IN HALF. ROLL THE TWO HALF EDGES

TOGETHER AND TAPE CLOSED.

13.   What 3D shape can you make from?   CIRCLE

14.   What 3D shape can you make from?   CUBE

15.   What 3D shape can you make from?   CYLINDER

16.   What 3D shape can you make from?   PYRAMID

142

17.   What 3D shape can you make from?   CUBOID

Match up the shapes with similar shapes from a different viewpoint.

18.   Draw a line from the name of the shape to the shape from different viewpoint.

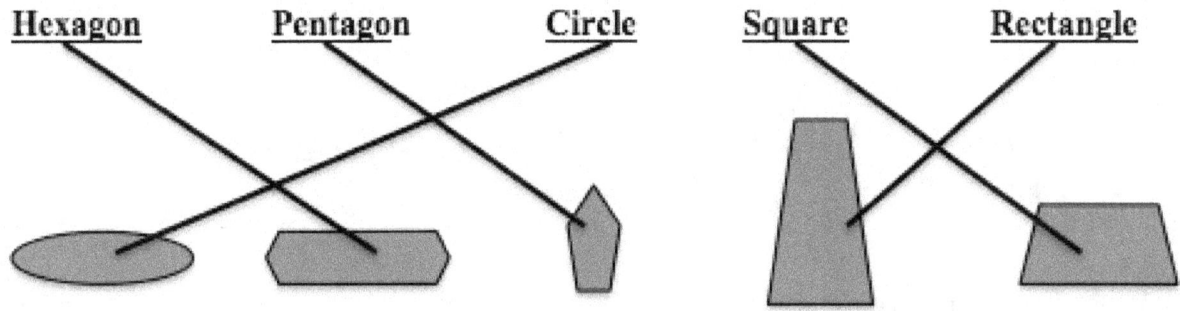

20.   What shape is the yellow portion of this figure?

   HEXAGON

   What shapes are the purple portions?

   PENTAGONS

Temperature Worksheet Answer Key

1. What is the coldest place on earth?   ☐ North Pole

   Check the correct answer.   ☐ **Antarctica**

                                           ☐ Canada

2. What is temperature? <u>TEMPERATURE IS HOW COLD OR HOT</u>

   <u>SOMETHING IS AT ANY GIVEN MOMENT IN TIME</u>.

3. What is a thermometer? <u>THERMOMETER IS MEASURES THE</u>

   <u>TEMPERATURE OF SOMETHING</u>.

Look at the two thermometers.

4. What is the temperature in degrees on the thermometer on the left side?

   <u>20°</u>

5. What is the temperature on the thermometer on the right side?

60°

6. According to this thermometer, it is 74°F.   What is the approximate temperature in Celsius?

23°C or 24°C

7. Is this a picture of an analog or digital thermometer?

ANALOG

8. What is the symbol for degrees?   °

9. What is the symbol for Fahrenheit?   F

10. What is the symbol for Celsius?   C

11. Using the correct symbols, how would you write the temperature for 65

degrees Fahrenheit?

<u>65°F</u>

12. Using the correct symbols, how would you write the temperature for 30 degrees Celsius?

<u>30°C</u>

13. At what Fahrenheit temperature does water boil? Check the correct answer.

☐   150° F

☐   **212° F**

☐   250° F

13. At what Celsius temperature does water boil?

☐   300° C

☐   500° C

☐   **100° C**

14. What is the average normal body temperature in Fahrenheit?

☐   96.8° F

☐   97.0° F

☐   **98.6° F**

15. What is the average normal body temperature in Celsius?

☐   **37.0° C**

☐   36.6° C

☐   35.8° C

16. What is the hottest time of the day?   MID-DAY

17. How does the sun affect temperatures? _____

IT USUALY MAKES THE TEMPERATURES WARMER.

18. How does rain affect temperatures and why? _____

    IT USUALLY MAKES THE TEMPERATURES COOLER.

    _____

19. Mark this thermometer to show 70° F.

20. What is the highest temperature measurement available on this thermometer?

    120°

    What is the lowest temperature measurement available on this thermometer?

    -40°

Time Worksheet Answer Key

1. What year is it now?   2013

    What year was it last year?   2012

    What year will it be next year? 2014

    What year were you born?   _____

2. What does chronological mean? _____

    <ins>IN ORDER OF TIME</ins>

3. Draw a line from the months of the year to their chronological order in the year.

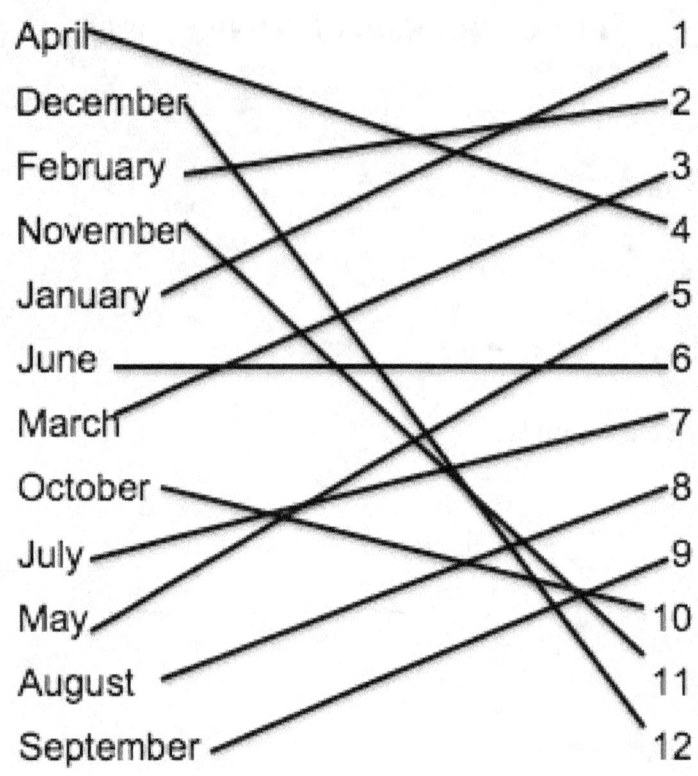

April	1
December	2
February	3
November	4
January	5
June	6
March	7
October	8
July	9
May	10
August	11
September	12

4. How many weeks are in a year?   <u>52</u>

5. What does PM mean?   Check the correct answer.

☐ **Evening**

☐ Noon

☐ Morning

6. How many days are in a week?   <u>7</u>

7. Draw a line from the days of the week to their chronological order in the week.

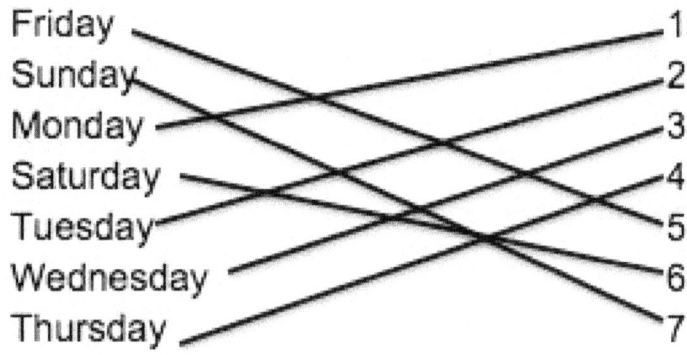

8. How many days are in a year? 365

Look at the calendar below and answer the following questions:

9. On what day is the 25th of January? MONDAY

10. What is the date of the second Monday in January? 11

11. How many Sundays are in January? 5

12. What day is New Year's Day? FRIDAY

13. What day is the last day of the month? SUNDAY

14. What day of the month is the 20th?   WEDNESDAY

15. How many days are in the month of January?   31

16. What time is it on this clock?

7:00

To what number is the hour hand pointing?

7

To what number is the minute hand pointing?

12

17. What is the time on this clock?

2:50

18. What time is it on this clock?

8:00

19. What is the time on this clock?

9:20

20. What time is it on this clock?

Using Shapes Worksheet Answer Key

In the graphic below, color the shapes as indicated:

1.   Color all of the squares red.

2.   Color all of the rectangles blue.

3.   Color all of the circles yellow.

4.   Color all of the triangles in green.

5.   Choose items from the list below that can be made with square shapes. Check all that apply.

☐ sphere   ☐ **cube**   ☐ **checker board**

☐ cow   ☐ **house**   ☐ **quilt**

☐ tree   ☑ fish   ☐ **chair**

☐ bucket   ☐ car   ☐ airplane

6.   Name four items that can be made by using rectangle shapes.

There are many possible answers like: Buildings, Notebook Paper, Bookcases, Televisions, Sofas, Little Red Wagons, Wood Planks, Books, Photographs, Bricks, Gift Boxes, Pans, etc.

7.   Look in a magazine and find four objects that have circle shapes. Paste them on the back of this paper.

8.   Cut out six identical squares. Using tape, assemble them into a cube.

9.   Cut out six shapes: Four identical rectangles and 2 identical squares. Using tape, assemble them into a cuboid.

10.   Cut out a circle. Fold it in half and then cut along the fold to make two halves. Take one half circle and fold the straight edge in half. Bring the two edges of the half circle together to form a cone. Tape the edges together.

11.   Cut out four identical triangles. Measure the base of the triangles. Cut out a square with sides that are the same length as the base of the triangles. Using tape, assemble them into a four-sided pyramid.

12.   Look in a magazine and find six objects that have triangle shapes. Paste them on the back of this paper.

13.   Make a triangle fish. Cut out two triangles that are the same size from two different colors of paper. Paste one triangle on top of the point of the other triangle. Using a marker, add an eye and a mouth. Decorate the fish with glitter and markers. Mount your finished fish on a sheet of blue paper and draw waves for the fish to swim in and bubbles coming up from his mouth.

14. Make a bull's eye with circles. Cut out five sizes of circles, each in a different color of paper. Paste them on top of each other in order of size with the largest circle on the bottom the smallest circle on top. Make sure each circle is centered on top of the next circle. Mount your bull's eye on a sheet of colorful construction paper

and hang it on your wall.

15. Make a snowman from hexagons. Cut out three different sizes of hexagons in white construction paper. Paste the three hexagons stacked on top of each other with the largest one on the bottom and the smallest one on the top. Add a hat in the shape of a triangle and two little round eyes. Paste on a heart shaped mouth. Add three round buttons glued on the second hexagon along with two straight line arms. Color some background for your snowman to enjoy! Pattern pieces are on the last page of this lesson.

16. What is the most famous pentagon

building in the world?

THE PENTAGON IN WASHINGTON, D.C.

17. What shape makes up the design on a soccer ball?

PENTAGON

18. Draw a dinosaur and put scales on his back using pentagons. You can

color them or cut them out with construction paper and glue them onto the back of the dinosaur. Use this dinosaur as a guide. You can even make baby dinosaurs!

19.   Name the four shapes in this graphic:

SQUARE

PENTAGON

TRIANGLE

CIRCLE

20.   How many sides does a hexagon have?

6

How many sides does a pentagon have?

5

Weight Worksheet Answer Sheet

Fill in the blanks for questions 1 through 3:

1. <u>WEIGHT</u>   is how heavy something is.

2. <u>MASS</u>   is how much space something occupies.

3. <u>MATTER</u>   makes up mass and can be heavy or light.

4.   When you make a guess or prediction, it is called

☐ an observation.

☐ **an hypotheses.**

☐ a question.

5.   If you want to weigh heavy objects, you need what type of scale?

☐ kitchen scale

☐ balance scale

☐ **bathroom scale**

6. If you want to weigh light weight items, you need what type of scale?

   ☐ balance scale

   ☐ bathroom scale

   ☐ **kitchen scale**

7. Step on the bathroom scale. How much do you weigh? _____

8. Using the balance scale, which item was heavier?

   ☐ **marble**

   ☐ cotton ball

   ☐ Styrofoam ball

9. When you are weighing two items on the balance scale and the beam of the scale does not dip, what does that mean?

   IT MEANS THE ITEMS ARE THE SAME WEIGHT.

10. Circle the object that is heavier? ELEPHANT

11. Circle the object that is lighter. BICYCLE

12. Circle the object that is heavier. SCISSORS

13. Circle the object that is lighter: FLOWER

14. Circle the object that is heaviest: WHALE

15. Circle the object that is lightest: BUTTERFLY

16. Number items in order from lightest to heaviest. 1 is for lightest (FISH), 2 is middle weight (MONKEY), & 3 is for the heaviest (ELEPHANT).

17. Number items in order from heaviest to lightest. 1 is for heaviest (BUILDING), 2 is middle weight (BUS), & 3 is for lightest (TRICYCLE).

18. Number items in order from lightest to heaviest. 1 is for lightest (FOOTBALL), 2 is middle weight (MOTORCYCLE), & 3 is for the heaviest (FIRE TRUCK).

19. Number items in order from heaviest to lightest. 1 is for heaviest (CAR), 2 is middle weight (BICYCLE), and 3 is for lightest (SHOES).

20. Circle the balancing scale. THE ONE ON THE LEFT

www.ingramcontent.com/pod-product-compliance
Lightning Source LLC
Chambersburg PA
CBHW081450170526

45166CB00008B/2381